OFFICIAL SQA PAST PAPERS WITH ANSWERS

ADVANCED HIGHER

MATHEMATICS
2008-2012

First exam published in 2008.
Published by Bright Red Publishing Ltd, 6 Stafford Street, Edinburgh EH3 7AU
tel: 0131 220 5804 fax: 0131 220 6710 info@brightredpublishing.co.uk www.brightredpublishing.co.uk

ISBN 978-1-84948-303-2

A CIP Catalogue record for this book is available from the British Library.

Bright Red Publishing is grateful to the copyright holders, as credited on the final page of the Question Section, for permission to use their material. Every effort has been made to trace the copyright holders and to obtain their permission for the use of copyright material. Bright Red Publishing will be happy to receive information allowing us to rectify any error or omission in future editions.

[BLANK PAGE]

X100/701

NATIONAL
QUALIFICATIONS
2008

TUESDAY, 20 MAY
1.00 PM – 4.00 PM

MATHEMATICS
ADVANCED HIGHER

Read carefully

1. Calculators may be used in this paper.

2. Candidates should answer **all** questions.

3. **Full credit will be given only where the solution contains appropriate working.**

Marks

Answer all the questions.

1. The first term of an arithmetic sequence is 2 and the 20th term is 97. Obtain the sum of the first 50 terms.   **4**

2. (a) Differentiate $f(x) = \cos^{-1}(3x)$ where $-\frac{1}{3} < x < \frac{1}{3}$.   **2**

   (b) Given $x = 2\sec\theta$, $y = 3\sin\theta$, use parametric differentiation to find $\dfrac{dy}{dx}$ in terms of θ.   **3**

3. Part of the graph $y = f(x)$ is shown below, where the dotted lines indicate asymptotes. Sketch the graph $y = -f(x+1)$ showing its asymptotes. Write down the equations of the asymptotes.   **4**

4. Express $\dfrac{12x^2 + 20}{x(x^2 + 5)}$ in partial fractions.   **3**

   Hence evaluate

   $$\int_1^2 \frac{12x^2 + 20}{x(x^2 + 5)}\, dx.$$ **3**

5. A curve is defined by the equation $xy^2 + 3x^2y = 4$ for $x > 0$ and $y > 0$.

   Use implicit differentiation to find $\dfrac{dy}{dx}$.   **3**

   Hence find an equation of the tangent to the curve where $x = 1$.   **3**

Marks

6. Let the matrix $A = \begin{pmatrix} 1 & x \\ x & 4 \end{pmatrix}$.

    (*a*)   Obtain the value(s) of x for which A is singular.   **2**

    (*b*)   When $x = 2$, show that $A^2 = pA$ for some constant p.

         Determine the value of q such that $A^4 = qA$.   **3**

7. Use integration by parts to obtain $\int 8x^2 \sin 4x \, dx$.   **5**

8. Write down and simplify the general term in the expansion of $\left(x^2 + \dfrac{1}{x}\right)^{10}$.   **3**

     Hence, or otherwise, obtain the term in x^{14}.   **2**

9. Write down the derivative of $\tan x$.   **1**

     Show that $1 + \tan^2 x = \sec^2 x$.   **1**

     Hence obtain $\int \tan^2 x \, dx$.   **2**

10. A body moves along a straight line with velocity $v = t^3 - 12t^2 + 32t$ at time t.

    (*a*)   Obtain the value of its acceleration when $t = 0$.   **1**

    (*b*)   At time $t = 0$, the body is at the origin O. Obtain a formula for the displacement of the body at time t.   **2**

         Show that the body returns to O, and obtain the time, T, when this happens.   **2**

11. For each of the following statements, decide whether it is true or false and prove your conclusion.

     A   For all natural numbers m, if m^2 is divisible by 4 then m is divisible by 4.

     B   The cube of any odd integer p plus the square of any even integer q is always odd.   **5**

12. Obtain the first three non-zero terms in the Maclaurin expansion of $x \ln(2 + x)$.   **3**

     Hence, or otherwise, deduce the first three non-zero terms in the Maclaurin expansion of $x \ln(2 - x)$.   **2**

     Hence obtain the first **two** non-zero terms in the Maclaurin expansion of $x \ln(4 - x^2)$.   **2**

     [*Throughout this question, it can be assumed that* $-2 < x < 2$.]

[Turn over for Questions 13 to 16 on *Page four*

Marks

13. Obtain the general solution of the differential equation

$$\frac{d^2 y}{dx^2} - 3\frac{dy}{dx} + 2y = 2x^2.$$ 7

Given that $y = \frac{1}{2}$ and $\frac{dy}{dx} = 1$, when $x = 0$, find the particular solution.   3

14. (*a*) Find an equation of the plane π_1 through the points $A(1, 1, 1)$, $B(2, -1, 1)$ and $C(0, 3, 3)$.   3

(*b*) The plane π_2 has equation $x + 3y - z = 2$.

Given that the point $(0, a, b)$ lies on both the planes π_1 and π_2, find the values of a and b. Hence find an equation of the line of intersection of the planes π_1 and π_2.   4

(*c*) Find the size of the acute angle between the planes π_1 and π_2.   3

15. Let $f(x) = \dfrac{x}{\ln x}$ for $x > 1$.

(*a*) Derive expressions for $f'(x)$ and $f''(x)$, simplifying your answers.   2,2

(*b*) Obtain the coordinates and nature of the stationary point of the curve $y = f(x)$.   3

(*c*) Obtain the coordinates of the point of inflexion.   2

16. Given $z = \cos\theta + i\sin\theta$, use de Moivre's theorem to write down an expression for z^k in terms of θ, where k is a positive integer.

Hence show that $\dfrac{1}{z^k} = \cos k\theta - i\sin k\theta.$ 3

Deduce expressions for $\cos k\theta$ and $\sin k\theta$ in terms of z.   2

Show that $\cos^2\theta \ \sin^2\theta = -\dfrac{1}{16}\left(z^2 - \dfrac{1}{z^2}\right)^2.$ 3

Hence show that $\cos^2\theta \ \sin^2\theta = a + b\cos 4\theta$, for suitable constants a and b.   2

[END OF QUESTION PAPER]

ADVANCED HIGHER

2009

[BLANK PAGE]

X100/701

NATIONAL
QUALIFICATIONS
2009

THURSDAY, 21 MAY
1.00 PM – 4.00 PM

MATHEMATICS
ADVANCED HIGHER

Read carefully

1. Calculators may be used in this paper.

2. Candidates should answer **all** questions.

3. **Full credit will be given only where the solution contains appropriate working.**

Marks

Answer all the questions.

1. (a) Given $f(x) = (x + 1)(x - 2)^3$, obtain the values of x for which $f'(x) = 0$.   **3**

   (b) Calculate the gradient of the curve defined by $\dfrac{x^2}{y} + x = y - 5$ at the point $(3, -1)$.   **4**

2. Given the matrix $A = \begin{pmatrix} t+4 & 3t \\ 3 & 5 \end{pmatrix}$.

   (a) Find A^{-1} in terms of t when A is non-singular.   **3**

   (b) Write down the value of t such that A is singular.   **1**

   (c) Given that the transpose of A is $\begin{pmatrix} 6 & 3 \\ 6 & 5 \end{pmatrix}$, find t.   **1**

3. Given that

$$x^2 e^y \frac{dy}{dx} = 1$$

   and $y = 0$ when $x = 1$, find y in terms of x.   **4**

4. Prove by induction that, for all positive integers n,

$$\sum_{r=1}^{n} \frac{1}{r(r+1)} = 1 - \frac{1}{n+1}.$$ **5**

5. Show that

$$\int_{\ln \frac{3}{2}}^{\ln 2} \frac{e^x + e^{-x}}{e^x - e^{-x}} \, dx = \ln \frac{9}{5}.$$ **4**

6. Express $z = \dfrac{(1+2i)^2}{7-i}$ in the form $a + ib$ where a and b are real numbers.

   Show z on an Argand diagram and evaluate $|z|$ and arg (z).   **6**

Marks

7. Use the substitution $x = 2\sin\theta$ to obtain the exact value of $\displaystyle\int_0^{\sqrt{2}} \frac{x^2}{\sqrt{4-x^2}}\,dx$.

   (Note that $\cos 2A = 1 - 2\sin^2 A$.)   **6**

8.   (a)   Write down the binomial expansion of $(1 + x)^5$.   **1**

     (b)   Hence show that $0 \cdot 9^5$ is $0 \cdot 59049$.   **2**

9. Use integration by parts to obtain the exact value of $\displaystyle\int_0^1 x\tan^{-1}x^2\,dx$.   **5**

10. Use the Euclidean algorithm to obtain the greatest common divisor of 1326 and 14654, expressing it in the form $1326a + 14654b$, where a and b are integers.   **4**

11. The curve $y = x^{2x^2+1}$ is defined for $x > 0$. Obtain the values of y and $\dfrac{dy}{dx}$ at the point where $x = 1$.   **5**

12. The first two terms of a geometric sequence are $a_1 = p$ and $a_2 = p^2$. Obtain expressions for S_n and S_{2n} in terms of p, where $S_k = \displaystyle\sum_{j=1}^{k} a_j$.   **1,1**

    Given that $S_{2n} = 65 S_n$ show that $p^n = 64$.   **2**

    Given also that $a_3 = 2p$ and that $p > 0$, obtain the exact value of p and hence the value of n.   **1,1**

13. The function $f(x)$ is defined by

    $$f(x) = \frac{x^2 + 2x}{x^2 - 1} \qquad (x \neq \pm 1).$$

    Obtain equations for the asymptotes of the graph of $f(x)$.   **3**

    Show that $f(x)$ is a strictly decreasing function.   **3**

    Find the coordinates of the points where the graph of $f(x)$ crosses
    (i) the x-axis and
    (ii) the horizontal asymptote.   **2**

    Sketch the graph of $f(x)$, showing clearly all relevant features.   **2**

[Turn over for Questions 14 to 16 on *Page four*

Marks

14. Express $\dfrac{x^2 + 6x - 4}{(x+2)^2(x-4)}$ in partial fractions.   **4**

Hence, or otherwise, obtain the first three non-zero terms in the Maclaurin

expansion of $\dfrac{x^2 + 6x - 4}{(x+2)^2(x-4)}$.   **5**

15.   (*a*)   Solve the differential equation

$$(x+1)\frac{dy}{dx} - 3y = (x+1)^4$$

given that $y = 16$ when $x = 1$, expressing the answer in the form $y = f(x)$.   **6**

   (*b*)   Hence find the area enclosed by the graphs of $y = f(x)$, $y = (1 - x)^4$ and the
x-axis.   **4**

16.   (*a*)   Use Gaussian elimination to solve the following system of equations

$$x + y - z = 6$$
$$2x - 3y + 2z = 2$$
$$-5x + 2y - 4z = 1.$$

      5

   (*b*)   Show that the line of intersection, L, of the planes $x + y - z = 6$ and
$2x - 3y + 2z = 2$ has parametric equations

$$x = \lambda$$
$$y = 4\lambda - 14$$
$$z = 5\lambda - 20.$$

      2

   (*c*)   Find the acute angle between line L and the plane $-5x + 2y - 4z = 1$.   **4**

[END OF QUESTION PAPER]

2010

[BLANK PAGE]

X100/701

NATIONAL
QUALIFICATIONS
2010

FRIDAY, 21 MAY
1.00 PM – 4.00 PM

MATHEMATICS
ADVANCED HIGHER

Read carefully

1. Calculators may be used in this paper.

2. Candidates should answer **all** questions.

3. **Full credit will be given only where the solution contains appropriate working.**

Marks

Answer all the questions.

1. Differentiate the following functions.

    (a) $f(x) = e^x \sin x^2$. **3**

    (b) $g(x) = \dfrac{x^3}{(1 + \tan x)}$. **3**

2. The second and third terms of a geometric series are −6 and 3 respectively. Explain why the series has a sum to infinity, and obtain this sum. **5**

3. (a) Use the substitution $t = x^4$ to obtain $\displaystyle\int \dfrac{x^3}{1 + x^8}\, dx$. **3**

    (b) Integrate $x^2 \ln x$ with respect to x. **4**

4. Obtain the 2×2 matrix M associated with an enlargement, scale factor 2, followed by a clockwise rotation of $60°$ about the origin. **4**

5. Show that
$$\binom{n+1}{3} - \binom{n}{3} = \binom{n}{2}$$

    where the integer n is greater than or equal to 3. **4**

6. Given $\mathbf{u} = -2\mathbf{i} + 5\mathbf{k}$, $\mathbf{v} = 3\mathbf{i} + 2\mathbf{j} - \mathbf{k}$ and $\mathbf{w} = -\mathbf{i} + \mathbf{j} + 4\mathbf{k}$.

    Calculate $\mathbf{u} \cdot (\mathbf{v} \times \mathbf{w})$. **4**

Marks

7. Evaluate

$$\int_1^2 \frac{3x+5}{(x+1)(x+2)(x+3)} \, dx$$

expressing your answer in the form $\ln \frac{a}{b}$, where a and b are integers.

6

8. (a) Prove that the product of two odd integers is odd.

2

(b) Let p be an odd integer. Use the result of (a) to prove by induction that p^n is odd for all positive integers n.

4

9. Obtain the first three non-zero terms in the Maclaurin expansion of $(1 + \sin^2 x)$.

4

10. The diagram below shows part of the graph of a function $f(x)$. State whether $f(x)$ is odd, even or neither. Fully justify your answer.

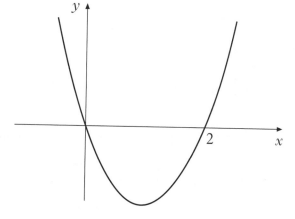

3

11. Obtain the general solution of the equation

$$\frac{d^2 y}{dx^2} + 4\frac{dy}{dx} + 5y = 0.$$

4

Hence obtain the solution for which $y = 3$ when $x = 0$ and $y = e^{-\pi}$ when $x = \frac{\pi}{2}$.

3

Marks

12. Prove by contradiction that if x is an irrational number, then $2 + x$ is irrational.   **4**

13. Given $y = t^3 - \dfrac{5}{2}t^2$ and $x = \sqrt{t}$ for $t > 0$, use parametric differentiation to express $\dfrac{dy}{dx}$ in terms of t in simplified form.   **4**

Show that $\dfrac{d^2 y}{dx^2} = at^2 + bt$, determining the values of the constants a and b.   **3**

Obtain an equation for the tangent to the curve which passes through the point of inflexion.   **3**

14. Use Gaussian elimination to show that the set of equations

$$x - y + z = 1$$
$$x + y + 2z = 0$$
$$2x - y + az = 2$$

has a unique solution when $a \neq 2{\cdot}5$.   **5**

Explain what happens when $a = 2{\cdot}5$.   **1**

Obtain the solution when $a = 3$.   **1**

Given $A = \begin{pmatrix} 5 & 2 & -3 \\ 1 & 1 & -1 \\ -3 & -1 & 2 \end{pmatrix}$ and $B = \begin{pmatrix} 1 \\ 0 \\ 2 \end{pmatrix}$, calculate AB.   **1**

Hence, or otherwise, state the relationship between A and the matrix

$$C = \begin{pmatrix} 1 & -1 & 1 \\ 1 & 1 & 2 \\ 2 & -1 & 3 \end{pmatrix}.$$ **2**

Marks

15. A new board game has been invented and the symmetrical design on the board is made from four identical "petal" shapes. One of these petals is the region enclosed between the curves $y = x^2$ and $y^2 = 8x$ as shown shaded in diagram 1 below.

Calculate the area of the complete design, as shown in diagram 2.

5

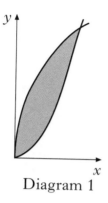

Diagram 1

Diagram 2

The counter used in the game is formed by rotating the shaded area shown in diagram 1 above, through $360°$ about the y-axis. Find the volume of plastic required to make one counter.

5

16. Given $z = r(\cos\theta + i\sin\theta)$, use de Moivre's theorem to express z^3 in polar form.

1

Hence obtain $\left(\cos\frac{2\pi}{3} + i\sin\frac{2\pi}{3}\right)^3$ in the form $a + ib$.

2

Hence, or otherwise, obtain the roots of the equation $z^3 = 8$ in Cartesian form.

4

Denoting the roots of $z^3 = 8$ by z_1, z_2, z_3:

(a) state the value $z_1 + z_2 + z_3$;

(b) obtain the value of $z_1^6 + z_2^6 + z_3^6$.

3

[*END OF QUESTION PAPER*]

[BLANK PAGE]

[BLANK PAGE]

X100/701

NATIONAL QUALIFICATIONS 2011	WEDNESDAY, 18 MAY 1.00 PM – 4.00 PM	MATHEMATICS ADVANCED HIGHER

Read carefully

1. Calculators may be used in this paper.

2. Candidates should answer **all** questions.

3. **Full credit will be given only where the solution contains appropriate working.**

Marks

Answer all the questions.

1. Express $\dfrac{13-x}{x^2+4x-5}$ in partial fractions and hence obtain

$$\int \frac{13-x}{x^2+4x-5}dx.$$

5

2. Use the binomial theorem to expand $\left(\dfrac{1}{2}x-3\right)^4$ and simplify your answer.

3

3. (a) Obtain $\dfrac{dy}{dx}$ when y is defined as a function of x by the equation

$$y+e^y=x^2.$$

3

(b) Given $f(x)=\sin x\cos^3 x$, obtain $f'(x)$.

Express $f'(x)$ in the form $\dfrac{g(x)f(x)}{\sin x\ \cos x}$.

3

4. (a) For what value of λ is $\begin{pmatrix} 1 & 2 & -1 \\ 3 & 0 & 2 \\ -1 & \lambda & 6 \end{pmatrix}$ singular?

3

(b) For $A=\begin{pmatrix} 2 & 2\alpha-\beta & -1 \\ 3\alpha+2\beta & 4 & 3 \\ -1 & 3 & 2 \end{pmatrix}$, obtain values of α and β such that

$$A'=\begin{pmatrix} 2 & -5 & -1 \\ -1 & 4 & 3 \\ -1 & 3 & 2 \end{pmatrix}.$$

3

5. Obtain the first four terms in the Maclaurin series of $\sqrt{1+x}$, and hence write down the first four terms in the Maclaurin series of $\sqrt{1+x^2}$.

4

Hence obtain the first four terms in the Maclaurin series of $\sqrt{(1+x)(1+x^2)}$.

2

Marks

6.

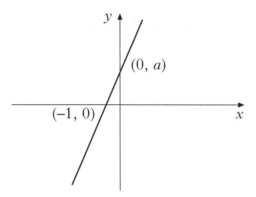

The diagram shows part of the graph of a function $f(x)$. Sketch the graph of $|f^{-1}(x)|$ showing the points of intersection with the axes.

4

7. A curve is defined by the equation $y = \dfrac{e^{\sin x}(2 + x)^3}{\sqrt{1 - x}}$ for $x < 1$.

Calculate the gradient of the curve when $x = 0$.

4

8. Write down an expression for $\displaystyle\sum_{r=1}^{n} r^3 - \left(\sum_{r=1}^{n} r \right)^2$

1

and an expression for

$$\sum_{r=1}^{n} r^3 + \left(\sum_{r=1}^{n} r \right)^2.$$

3

9. Given that $y > -1$ and $x > -1$, obtain the general solution of the differential equation

$$\frac{dy}{dx} = 3(1 + y)\sqrt{1 + x}$$

expressing your answer in the form $y = f(x)$.

5

[Turn over

Marks

10. Identify the locus in the complex plane given by

$$|z - 1| = 3.$$

Show in a diagram the region given by $|z - 1| \leq 3$.   5

11. (*a*) Obtain the exact value of $\int_0^{\pi/4} (\sec x - x)(\sec x + x)dx$.   3

(*b*) Find $\int \dfrac{x}{\sqrt{1 - 49x^4}}dx$.   4

12. Prove by induction that $8^n + 3^{n-2}$ is divisible by 5 **for all integers $n \geq 2$.**   5

13. The first three terms of an arithmetic sequence are $a, \dfrac{1}{a}, 1$ where $a < 0$.

Obtain the value of a and the common difference.   5

Obtain the smallest value of n for which the sum of the first n terms is greater than 1000.   4

14. Find the general solution of the differential equation

$$\frac{d^2 y}{dx^2} - \frac{dy}{dx} - 2y = e^x + 12.$$ 7

Find the particular solution for which $y = -\dfrac{3}{2}$ and $\dfrac{dy}{dx} = \dfrac{1}{2}$ when $x = 0$.   3

Marks

15. The lines L_1 and L_2 are given by the equations

$$\frac{x-1}{k} = \frac{y}{-1} = \frac{z+3}{1} \text{ and } \frac{x-4}{1} = \frac{y+3}{1} = \frac{z+3}{2},$$

respectively.

Find:

(a)   the value of k for which L_1 and L_2 intersect and the point of intersection;   **6**

(b)   the acute angle between L_1 and L_2.   **4**

16. Define $I_n = \displaystyle\int_0^1 \frac{1}{(1+x^2)^n}\,dx$ for $n \geq 1$.

(a)   Use integration by parts to show that

$$I_n = \frac{1}{2^n} + 2n\int_0^1 \frac{x^2}{(1+x^2)^{n+1}}\,dx.$$ **3**

(b)   Find the values of A and B for which

$$\frac{A}{(1+x^2)^n} + \frac{B}{(1+x^2)^{n+1}} = \frac{x^2}{(1+x^2)^{n+1}}$$

and hence show that

$$I_{n+1} = \frac{1}{n\times 2^{n+1}} + \left(\frac{2n-1}{2n}\right)I_n.$$ **5**

(c)   Hence obtain the exact value of $\displaystyle\int_0^1 \frac{1}{(1+x^2)^3}\,dx.$ **3**

[END OF QUESTION PAPER]

[BLANK PAGE]

2012

[BLANK PAGE]

X100/13/01

NATIONAL   MONDAY, 21 MAY   MATHEMATICS
QUALIFICATIONS   1.00 PM – 4.00 PM   ADVANCED HIGHER
2012

Read carefully

1   Calculators may be used in this paper.

2   Candidates should answer **all** questions.

3   **Full credit will be given only where the solution contains appropriate working.**

Marks

Answer all the questions

1. (*a*) Given $f(x) = \dfrac{3x+1}{x^2+1}$, obtain $f'(x)$.   **3**

    (*b*) Let $g(x) = \cos^2 x \exp(\tan x)$. Obtain an expression for $g'(x)$ and simplify your answer.   **4**

2. The first and fourth terms of a geometric series are 2048 and 256 respectively. Calculate the value of the common ratio.   **2**

    Given that the sum of the first n terms is 4088, find the value of n.   **3**

3. Given that $(-1 + 2i)$ is a root of the equation
    $$z^3 + 5z^2 + 11z + 15 = 0,$$
    obtain all the roots.   **4**

    Plot all the roots on an Argand diagram.   **2**

4. Write down and simplify the general term in the expansion of $\left(2x - \dfrac{1}{x^2}\right)^9$.   **3**

    Hence, or otherwise, obtain the term independent of x.   **2**

5. Obtain an equation for the plane passing through the points $P(-2, 1, -1)$, $Q(1, 2, 3)$ and $R(3, 0, 1)$.   **5**

6. Write down the Maclaurin expansion of e^x as far as the term in x^3.   **1**

    Hence, or otherwise, obtain the Maclaurin expansion of $(1 + e^x)^2$ as far as the term in x^3.   **4**

7. A function is defined by $f(x) = |x + 2|$ for all x.

    (*a*) Sketch the graph of the function for $-3 \le x \le 3$.   **2**

    (*b*) On a separate diagram, sketch the graph of $f'(x)$.   **2**

8. Use the substitution $x = 4 \sin \theta$ to evaluate $\displaystyle\int_0^2 \sqrt{16 - x^2}\, dx$.   **6**

Marks

9. A non-singular $n \times n$ matrix A satisfies the equation $A + A^{-1} = I$, where I is the $n \times n$ identity matrix. Show that $A^3 = kI$ and state the value of k.

4

10. Use the division algorithm to express 1234_{10} in base 7.

3

11. (a) Write down the derivative of $\sin^{-1}x$.

1

    (b) Use integration by parts to obtain $\displaystyle\int \sin^{-1}x \cdot \frac{x}{\sqrt{1-x^2}}\,dx$.

4

12. The radius of a cylindrical column of liquid is decreasing at the rate of $0\cdot02$ m s^{-1}, while the height is increasing at the rate of $0\cdot01$ m s^{-1}.

Find the rate of change of the volume when the radius is $0\cdot6$ metres and the height is 2 metres.

5

[*Recall that the volume of a cylinder is given by $V = \pi r^2 h$.*]

13. A curve is defined parametrically, for all t, by the equations

$$x = 2t + \frac{1}{2}t^2, \qquad y = \frac{1}{3}t^3 - 3t.$$

Obtain $\dfrac{dy}{dx}$ and $\dfrac{d^2y}{dx^2}$ as functions of t.

5

Find the values of t at which the curve has stationary points and determine their nature.

3

Show that the curve has exactly two points of inflexion.

2

14. (a) Use Gaussian elimination to obtain the solution of the following system of equations in terms of the parameter λ.

$$4x + 6z = 1$$
$$2x - 2y + 4z = -1$$
$$-x + y + \lambda z = 2$$

5

    (b) Describe what happens when $\lambda = -2$.

1

    (c) When $\lambda = -1\cdot9$ the solution is $x = -22\cdot25$, $y = 8\cdot25$, $z = 15$.

       Find the solution when $\lambda = -2\cdot1$.

2

       Comment on these solutions.

1

[Turn over for Questions 15 and 16 on *Page four*

15. (*a*) Express $\dfrac{1}{(x-1)(x+2)^2}$ in partial fractions.   **4**

(*b*) Obtain the general solution of the differential equation

$$(x-1)\dfrac{dy}{dx} - y = \dfrac{x-1}{(x+2)^2},$$

expressing your answer in the form $y = f(x)$.   **7**

16. (*a*) Prove by induction that

$$(\cos\theta + i\sin\theta)^n = \cos n\theta + i\sin n\theta$$

for all integers $n \geq 1$.   **6**

(*b*) Show that the real part of $\dfrac{\left(\cos\dfrac{\pi}{18}+i\sin\dfrac{\pi}{18}\right)^{11}}{\left(\cos\dfrac{\pi}{36}+i\sin\dfrac{\pi}{36}\right)^{4}}$ is zero.   **4**

[END OF QUESTION PAPER]

ADVANCED HIGHER | ANSWER SECTION

SQA ADVANCED HIGHER MATHEMATICS 2008–2012

1. 6225

2. (a) $f'(x) = \dfrac{-3}{\sqrt{1-9x^2}}$

(b) $\dfrac{dy}{dx} = \dfrac{3\cos^3\theta}{2\sin\theta}$

3.

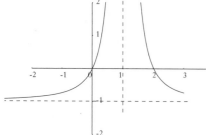

Asymptotes are $y = -1$ and $x = 1$.

4. $\dfrac{12x^2+20}{x(x^2+5)} = \dfrac{4}{x} + \dfrac{8x}{x^2+5}$

$\displaystyle\int_1^2 \dfrac{12x^2+20}{x(x^2+5)}\,dx = 4\ln 3$

5. $\dfrac{dy}{dx} = \dfrac{-y^2-6xy}{2xy+3x^2}$

$\dfrac{dy}{dx} = \dfrac{-7}{5}$

At $(1,1)$

Tangent is $5y + 7x = 12$

6. (a) $x = \pm 2$

(b) When $x = 2$, $A = \begin{pmatrix} 1 & 2 \\ 2 & 4 \end{pmatrix}$

$A^2 = \begin{pmatrix} 1 & 2 \\ 2 & 4 \end{pmatrix}\begin{pmatrix} 1 & 2 \\ 2 & 4 \end{pmatrix} = \begin{pmatrix} 5 & 10 \\ 10 & 20 \end{pmatrix} = 5A$

$A^4 = (A^2)^2 = (5A)^2 = 25A^2 = 125A$

7. $\int 8x^2 \sin 4x\,dx = -2x^2\cos 4x + x\sin 4x + \frac{1}{4}\cos 4x + c$

8. rth term is $\dbinom{10}{r}(x^2)^{10-r}\left(\dfrac{1}{x}\right)^r = \dbinom{10}{r}x^{20-3r}$

Requested term is $45x^{14}$

9. $\dfrac{d}{dx}(\tan x) = \sec^2 x$

$1 + \tan^2 x = 1 + \dfrac{\sin^2 x}{\cos^2 x}$

$= \dfrac{\cos^2 x + \sin^2 x}{\cos^2 x} = \sec^2 x.$

$\int \tan^2 x\,dx = \int (\sec^2 x - 1)\,dx = \tan x - x + c$

10. (a) $x''(t) = 3t^2 - 24t + 32$ so $x''(0) = 32$

(b) $(x)t = \frac{1}{4}$ (as $x(0) = 0$ so $c = 0$)

At O, $x = 0$

$\frac{1}{4}t^4 - 4t^3 + 16t^2 = 0$

$t^2(t^2 - 16t + 64) = 0$

$t^2(t-8)^2 = 0$

The body returns to O when $t = 8$.

11. A Counter example $m = 2$. So statement is false.

B Let the numbers be $2n + 1$ and $2m$.

Then $(2n+1)^3 + (2m)^2$

$= 8n^3 + 12n^2 + 6n + 1 + 4m^2$

$= 2(4n^3 + 6n^2 + 3n + 2m^2) + 1$ which is odd.

or, prove algebraically that either the cube of an odd number is odd or the square of an even number is even.

Odd cubed is odd and even squared is even. So the sum of them is odd.

12. $x\ln(2+x) = (\ln 2)x + \dfrac{x^2}{2} - \dfrac{x^3}{8} + \dots$

$x\ln(2-x) = (\ln 2)x - \dfrac{x^2}{2} - \dfrac{x^3}{8} + \dots$

$x\ln(4 - x^2) = (2\ln 2)x - \dfrac{x^3}{4} + \dots$

13. Complementary function: $y = Ae^x + Be^{2x}$

General solution: $y = Ae^x + Be^{2x} + x^2 + 3x + \dfrac{7}{2}$

Particular solution: $y = -4e^x + e^{2x} + x^2 + 3x + \dfrac{7}{2}$

14. (a) $2x + y = 3$

(b) $a = 3$, $b = 7$

$x = 0 + 2t$; $y = 3 - 4t$; $z = 7 - 10t$

(c) $\cos\theta = \dfrac{|\mathbf{a}\cdot\mathbf{b}|}{|\mathbf{a}||\mathbf{b}|} = \dfrac{5}{\sqrt{55}}$

$\left\{\text{or }\sin\theta = \dfrac{|\mathbf{a}\times\mathbf{b}|}{|\mathbf{a}||\mathbf{b}|} = \sqrt{\dfrac{6}{11}}\right\}$

hence $\theta \approx 47.6°$

15. (a) $f'(x) = \dfrac{\ln x - 1}{(\ln x)^2}$

$f''(x) = \dfrac{2 - \ln x}{x(\ln x)^3}$

(b) The point (e, e) is a minimum turning point.

(c) $x = e^2 \Rightarrow y = \dfrac{1}{2}e^2$

16. $z^k = \cos k\theta + i \sin k\theta$

so $\dfrac{1}{z^k} = \dfrac{1}{\cos k\theta + i \sin k\theta}$

$= \dfrac{\cos k\theta - i \sin k\theta}{\cos^2 k\theta + \sin^2 k\theta}$

$= \cos k\theta - i \sin k\theta.$

$\cos k\theta = \dfrac{1}{2}(z^k + z^{-k}) \quad \sin k\theta = \dfrac{1}{2i}(z^k + z^{-k})$

For $k = 1$

$\cos^2 \theta \sin^2 \theta = (\cos\theta \sin\theta)^2$

$= \left(\dfrac{(z + \frac{1}{z})(z - \frac{1}{z})}{4i} \right)^2$

$= -\dfrac{1}{16}\left(z^2 - \dfrac{1}{z^2} \right)^2.$

$\left(z^2 - \dfrac{1}{z^2} \right)^2 = z^4 + \dfrac{1}{z^4} - 2 = 2\cos 4\theta - 2$

$\Rightarrow \cos^2\theta \sin^2\theta = \dfrac{1}{8} - \dfrac{1}{8}\cos 4\theta,$

i.e. $a = \dfrac{1}{8}$ and $b = \dfrac{1}{8}$

ADVANCED HIGHER MATHEMATICS 2009

1. (a) $f(x) = (x + 1)(x - 2)^3$

$\begin{aligned}
f'(x) &= (x - 2)^3 + 3(x + 1)(x - 2)^2 \\
&= (x - 2)^2 ((x - 2) + 3(x + 1)) \\
&= (x - 2)^2 (4x + 1) \\
&= 0 \text{ when } x = 2 \text{ and when } x = -\dfrac{1}{4}.
\end{aligned}$

(b) $\dfrac{x^2}{y} + x = y - 5 \Rightarrow x^2 + xy = y^2 - 5y$

$2x + x\dfrac{dy}{dx} + y = 2y\dfrac{dy}{dx} - 5\dfrac{dy}{dx}$

$x = 3, y = -1 \Rightarrow 6 + 3\dfrac{dy}{dx} - 1 = -2\dfrac{dy}{dx} - 5\dfrac{dy}{dx}$

$5 = -10\dfrac{dy}{dx} \Rightarrow \boldsymbol{\dfrac{dy}{dx}} = \dfrac{-1}{2}$

2. (a) $\det\begin{pmatrix} t + 4 & 3t \\ 3 & 5 \end{pmatrix} = 5(t + 4) - 9t = 20 - 4t$

$A^{-1} = \dfrac{1}{20 - 4t}\begin{pmatrix} 5 & -3t \\ -3t & +4 \end{pmatrix}$

(b) $20 - 4t = 0 \Rightarrow t = 5$

(c) $\begin{pmatrix} t + 4 & 3 \\ 3t & 5 \end{pmatrix} = \begin{pmatrix} 6 & 3 \\ 6 & 5 \end{pmatrix} \Rightarrow t = 2$

3.

$e^y x^2 \dfrac{dy}{dx} = 1$

$e^y \dfrac{dy}{dx} = x^{-2}$

$\int e^y\, dy = \int x^{-2}\, dx$

$e^y = -x^{-1} + c$

$y = 0$ when $x = 1$ so $1 = -1 + c \Rightarrow c = 2$

$e^y = 2 - \dfrac{1}{x} \Rightarrow y = \ln\left(2 - \dfrac{1}{x} \right)$

4. When $n = 1$, LHS $= \dfrac{1}{1 \times 2} = \dfrac{1}{2}$, RHS $= 1 - \dfrac{1}{2} = \dfrac{1}{2}$.
So true when $n = 1$.

Assume true for $n = k$, $\displaystyle\sum_{r=1}^{k} \dfrac{1}{r(r+1)} = 1 - \dfrac{1}{k+1}$.

Consider $n = k + 1$

$\displaystyle\sum_{r=1}^{k+1} \dfrac{1}{r(r+1)} = \sum_{r=1}^{k} \dfrac{1}{r(r+1)} + \dfrac{1}{(k+1)(k+2)}$

$= 1 - \dfrac{1}{k+1} + \dfrac{1}{(k+1)(k+2)}$

$= 1 - \dfrac{k+2-1}{(k+1)(k+2)} = 1 - \dfrac{k+1}{(k+1)((k+1)+1)}$

$= 1 - \dfrac{1}{((k+1)+1)}$

Thus, if true for $n = k$, statement is true for $n = k + 1$, and, since true for $n = 1$, true for all $n \geq 1$.

5.
$$\int_{\ln\frac{3}{2}}^{\ln 2}\frac{e^x+e^{-x}}{e^x-e^{-x}}\,dx$$

Let $u=e^x-e^{-x}$, then $du=(e^x+e^{-x})\,dx$.

When $x=\ln\frac{3}{2}, u=\frac{3}{2}-\frac{2}{3}=\frac{5}{6}$ and when

$x=\ln 2, u=2-\frac{1}{2}=\frac{3}{2}$.

$$\int_{\ln\frac{3}{2}}^{\ln 2}\frac{e^x+e^{-x}}{e^x-e^{-x}}\,dx \int_{5/6}^{3/2}\frac{du}{u}$$

$$=\big[\ln u\big]_{5/6}^{3/2}$$

$$=\ln\frac{3}{2}-\ln\frac{5}{6}=\ln\frac{9}{5}$$

6.
$$\frac{(1+2i)^2}{7-i}=\frac{1+4i-4}{7-i}$$

$$=\frac{-3+4i}{7-i}\times\frac{7+i}{7+i}$$

$$=\frac{(-3+4i)(7+i)}{50}$$

$$=-\frac{1}{2}+\frac{1}{2}i$$

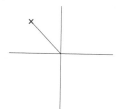

$$|z|=\sqrt{\frac{1}{4}+\frac{1}{4}}=\frac{1}{2}\sqrt{2}$$

$$\arg z=\tan^{-1}\frac{\frac{1}{2}}{-\frac{1}{2}}=\tan^{-1}(-1)=\frac{3\pi}{4}\text{ (or }135°).$$

7. $x=2\sin\theta\Rightarrow dx=2\cos\theta\,d\theta$

$x=0\Rightarrow\theta=0;\ x=\sqrt{2}\Rightarrow\sin\theta=\frac{1}{\sqrt{2}}\Rightarrow\theta=\frac{\pi}{4}$

$$\int_0^{\sqrt{2}}\frac{x^2}{\sqrt{4-x^2}}\,dx=\int_0^{\pi/4}\frac{4\sin^2\theta}{\sqrt{4-4\sin^2\theta}}(2\cos\theta)\,d\theta$$

$$=\int_0^{\pi/4}\frac{4\sin^2\theta}{2\cos\theta}(2\cos\theta)\,d\theta$$

$$=2\int_0^{\pi/4}(2\sin^2\theta)\,d\theta$$

$$=2\int_0^{\pi/4}(1-\cos2\theta)\,d\theta$$

$$=2\Big[\theta-\frac{1}{2}\sin2\theta\Big]_0^{\pi/4}$$

$$=2\Big\{\Big[\frac{\pi}{4}-\frac{1}{2}\Big]-0\Big\}$$

$$=\frac{\pi}{2}-1$$

8. (a) $(1+x)^5=1+5x+10x^2+10x^3+5x^4+x^5$

(b) Let $x=-0\cdot1$, then

$$0\cdot9^5=(1+(-0\cdot1))^5$$

$$=1-0\cdot5+0\cdot1-0\cdot01+0\cdot0005-0\cdot00001$$

$$=0\cdot5+0\cdot09+0\cdot00049$$

$$=0\cdot59049$$

9. $\displaystyle\int_0^1 x\tan^{-1}x^2\,dx=\Big[\tan^{-1}x^2\int x\,dx\Big]_0^1-\int_0^1\frac{2x}{1+x^4}\frac{x^2}{2}\,dx$

$$=\Big[\frac{1}{2}x^2\tan^{-1}x^2\Big]_0^1-\int_0^1\frac{x^3}{1+x^4}\,dx$$

$$=\Big[\frac{1}{2}x^2\tan^{-1}x^2\Big]_0^1-\Big[\frac{1}{4}\ln(1+x^4)\Big]_0^1$$

$$=\frac{1}{2}\tan^{-1}1-0-\Big[\frac{1}{4}\ln2-\frac{1}{4}\ln1\Big]$$

$$=\frac{\pi}{8}-\frac{1}{4}\ln2$$

10. $14654=11\times1326+68$

$$1326=19\times68+34$$

$$68=2\times34$$

$$34=1326-19\times68$$

$$=1326-19(14654-11\times1326)$$

$$=210\times1326-19\times14654$$

11. When $x=1, y=1$.

$$y=x^{2x^2+1}$$

$$\Rightarrow\ln y=\ln\left(x^{2x^2+1}\right)$$

$$=(2x^2+1)\ln x$$

$$\frac{1}{y}\frac{dy}{dx}=\frac{2x^2+1}{x}+4x\ln x$$

Hence, when $x=1, y=1$ and

$$\frac{dy}{dx}=3+0=3.$$

12. $a_j=p^j\Rightarrow S_k=p+p^2+\cdots+p^k=\dfrac{p\left(p^k-1\right)}{p-1}$

$$S_n=\frac{p\left(p^n-1\right)}{p-1}$$

$$S_{2n}=\frac{p\left(p^{2n}-1\right)}{p-1}$$

$$\frac{p\left(p^{2n}-1\right)}{p-1}=\frac{65p\left(p^n-1\right)}{p-1}$$

$$(p^n+1)(p^n-1)=65(p^n-1)$$

$$p^n+1=65$$

$$\Rightarrow p^n=64$$

$a_2=p^2\Rightarrow a_3=p^3$ but $a_3=2p$ so $p^3=2p$

$$\Rightarrow p^2=2\Rightarrow p=\sqrt{2}\ \text{ since }p>0.$$

$$p^n=64=2^6=\left(\sqrt{2}\right)^{12}$$

$$n=12$$

13.
$$f(x) = \frac{x^2 + 2x}{x^2 - 1} = \frac{x^2 + 2x}{(x-1)(x+1)}$$

Hence there are vertical asymptotes at $x = -1$ and $x = 1$.

$$f(x) = \frac{x^2 + 2x}{x^2 - 1} = \frac{1 + \dfrac{2x}{x^2}}{1 - \dfrac{1}{x^2}} = \frac{1 + \dfrac{2}{x}}{1 - \dfrac{1}{x^2}}$$

$$\to 1 \text{ as } x \to \infty.$$

So $y = 1$ is a horizontal asymptote.

$$f(x) = \frac{x^2 + 2x}{x^2 - 1}$$

$$f-(x) = \frac{(2x + 2)(x^2 - 1) - (x^2 + 2x)2x}{(x^2 - 1)^2}$$

$$= \frac{2x^3 - 2x + 2x^2 - 2 - 2x^3 - 4x^2}{(x^2 - 1)^2} = \frac{-2(x^2 + x + 1)}{(x^2 - 1)^2}$$

$$= \frac{-2\left(\left(x + \dfrac{1}{2}\right)^2 + \dfrac{3}{4}\right)}{(x^2 - 1)^2} < 0$$

Hence $f(x)$ is a strictly decreasing function.

$$f(x) = \frac{x^2 + 2x}{x^2 - 1} = 0 \Rightarrow x = 0 \text{ or } x = -2$$

$$f(x) = \frac{x^2 + 2x}{x^2 - 1} = 1 \Rightarrow x^2 + 2x = x^2 - 1 \Rightarrow x = -\frac{1}{2}$$

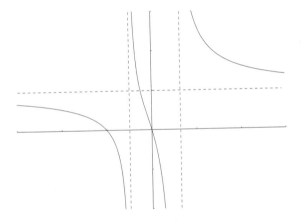

14.
$$\frac{x^2 + 6x - 4}{(x + 2)^2(x - 4)} = \frac{A}{(x + 2)^2} + \frac{B}{x + 2} + \frac{C}{x - 4}$$

$$x^2 + 6x - 4 = A(x - 4) + B(x + 2)(x - 4) + C(x + 2)^2$$

Let $x = -2$ then $4 - 12 - 4 = -6A \Rightarrow A = 2$.

Let $x = 4$ then $16 + 24 - 4 = 36C \Rightarrow C = 1$.

Let $x = 0$ then

$$-4 = -4A - 8B + 4C \Rightarrow -4 = -8 - 8B + 4 \Rightarrow B = 0.$$

Thus
$$\frac{x^2 + 6x - 4}{(x + 2)^2(x - 4)} = \frac{2}{(x + 2)^2} + \frac{1}{x - 4}.$$

Let $f(x) = 2(x + 2)^{-2} + (x - 4)^{-1}$ then

$$f(x) = 2(x + 2)^{-2} + (x - 4)^{-1} \Rightarrow f(0) = \frac{1}{2} - \frac{1}{4} = \frac{1}{4}$$

$$f'(x) = -4(x + 2)^{-3} - (x - 4)^{-2}$$

$$\Rightarrow f''(0) = \frac{1}{2} - \frac{1}{16} = -\frac{9}{16}$$

$$f''(x) = 12(x + 2)^{-4} + 2(x - 4)^{-3}$$

$$\Rightarrow f''(0) = \frac{3}{4} - \frac{1}{32} = \frac{23}{32}$$

Thus
$$\frac{x^2 + 6x - 4}{(x + 2)^2(x - 4)} = \frac{1}{4} - \frac{9x}{16} + \frac{23x^2}{64} + \dots$$

15. (a) $(x + 1)\dfrac{dy}{dx} - 3y = (x + 1)^4$

$$\frac{dy}{dx} - \frac{3}{x + 1}y = (x + 1)^3$$

Integrating factor:

Since $\displaystyle\int \frac{-3}{x + 1}dx = -3 \ln (x + 1)$.

Hence the integrating factor is $(x + 1)^{-3}$.

$$\frac{1}{(x + 1)^3}\frac{dy}{dx} - \frac{3}{(x + 1)^4}y = 1$$

$$\frac{d}{dx}\left((x + 1)^{-3}y\right) = 1$$

$$\frac{y}{(x + 1)^3} = \int 1 dx$$

$$= x + c$$

$y = 16$ when $x = 1$, so $2 = 1 + c \Rightarrow c = 1$. Hence

$$y = (x + 1)^4$$

(b) $(x + 1)^4 = (1 - x)^4$

$$x + 1 = 1 - x \Rightarrow x = 0$$

or $x + 1 = -1 + x$ which has no solutions.

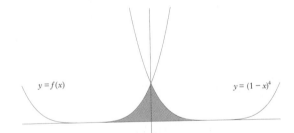

$$\text{Area} = \int_{-1}^{0}(x + 1)^4 dx + \int_{0}^{1}(1 - x)^4 dx$$

$$= 2\int_{-1}^{0}(x + 1)^4 dx$$

$$= \frac{2}{5}[(x + 1)^5]_{-1}^{0} = \frac{2}{5} - 0 = \frac{2}{5}$$

16. (*a*)
$$x + y - z = 6$$
$$2x - 3y + 2z = 2$$
$$-5x + 2y - 4z = 1$$

$$\begin{array}{ccc|c} 1 & 1 & -1 & 6 \\ 2 & -3 & 2 & 2 \\ -5 & 2 & -4 & 1 \end{array} \Rightarrow \begin{array}{ccc|c} 1 & 1 & -1 & 6 \\ 0 & -5 & 4 & -10 \\ 0 & 7 & -9 & 31 \end{array} \Rightarrow \begin{array}{ccc|c} 6 & 1 & 1 & -1 & 6 \\ -10 \Rightarrow 0 & -5 & 4 & -10 \\ 31 & 0 & 0 & -\frac{17}{5} & 17 \end{array}$$

$$z = 17 \div \left(\frac{-17}{5}\right) = -5$$

$$-5y - 20 = -10 \Rightarrow y = -2$$
$$x - 2 + 5 = 6 \Rightarrow x = 3$$

(*b*) Let $x = \lambda$.

In first plane: $x + y - z = 6$.

$\lambda + (4\lambda - 14) - (5\lambda - 20) = 5\lambda - 5\lambda + 6 = 6$.

In the second plane:

$2x - 3y + 2z = 2\lambda - 3(4\lambda - 14) + 2(5\lambda - 20) =$
$5\lambda - 5\lambda + 2 = 2$.

(*c*) Direction of L is $\mathbf{i} + 4\mathbf{j} + 5\mathbf{k}$, direction of normal to the plane is $-5\mathbf{i} + 2\mathbf{j} - 4\mathbf{k}$. Letting θ be the angle between these then

$$\cos\theta = \frac{-5 + 8 - 20}{\sqrt{42}\sqrt{45}}$$

$$= \frac{-17}{3\sqrt{210}}$$

This gives a value of $113 \cdot 0°$ which leads to the angle
$113 \cdot 0° - 90° = 23 \cdot 0°$.

ADVANCED HIGHER MATHEMATICS 2010

1. (*a*) For $f(x) = e^x \sin x^2$,
$$f'(x) = e^x \sin x^2 + e^x (2x \cos x^2).$$

(*b*) *Method 1*

For $g(x) = \dfrac{x^3}{(1 + \tan x)}$,
$$g'(x) = \frac{3x^2(1 + \tan x) - x^3 \sec^2 x}{(1 + \tan x)^2}.$$

Method 2
$$g(x) = x^3(1 + \tan x)^{-1}$$
$$g'(x) =$$
$$3x^2(1 + \tan x)^{-1} + x^3(-1)(1 + \tan x)^{-2}\sec^2 x$$
$$= \frac{x^2}{(1 + \tan x)^2}\left(3 + 3\tan x - x\sec^2 x\right)$$

2. Let the first term be a and the common ratio be r. Then
$$ar = -6 \quad \text{and} \quad ar^2 = 3$$
Hence
$$r = \frac{ar^2}{ar} = \frac{3}{-6} = -\frac{1}{2}.$$
So, since $|r| < 1$, the sum to infinity exists.
$$S = \frac{a}{1 - r}$$
$$= \frac{12}{1 - \left(-\frac{1}{2}\right)} = \frac{12}{\frac{3}{2}}$$
$$= 8.$$

3. (*a*)
$$t = x^4 \Rightarrow dt = 4x^3 dx$$
$$\int \frac{x^3}{1 + x^8} dx = \frac{1}{4}\int \frac{4x^3}{1 + (x^4)^2} dx$$
$$= \frac{1}{4}\int \frac{1}{1 + t^2} dt$$
$$= \frac{1}{4}\tan^{-1} t + c$$
$$= \frac{1}{4}\tan^{-1} x^4 + c$$

(*b*)
$$\int x^2 \ln x \, dx = \int (\ln x) x^2 dx$$
$$= \ln x \int x^2 dx - \int \frac{1}{x}\frac{x^3}{3} dx$$
$$= \frac{1}{3}x^3 \ln x - \frac{1}{3}\int x^2 dx$$
$$= \frac{1}{3}x^3 \ln x - \frac{1}{9}x^3 + c$$

4. The matrix $\begin{pmatrix} 2 & 0 \\ 0 & 2 \end{pmatrix}$ gives an enlargement, scale factor 2.

The matrix $\begin{pmatrix} \frac{1}{2} & \frac{\sqrt{3}}{2} \\ -\frac{\sqrt{3}}{2} & \frac{1}{2} \end{pmatrix}$ gives a clockwise rotation of $60°$ about the origin.

$$M = \begin{pmatrix} \frac{1}{2} & \frac{\sqrt{3}}{2} \\ -\frac{\sqrt{3}}{2} & \frac{1}{2} \end{pmatrix}\begin{pmatrix} 2 & 0 \\ 0 & 2 \end{pmatrix}$$
$$= \begin{pmatrix} 1 & \sqrt{3} \\ -\sqrt{3} & 1 \end{pmatrix}.$$

5.
$$\binom{n+1}{3} - \binom{n}{3} = \frac{(n+1)!}{3!\,(n-2)!} - \frac{n!}{3!\,(n-3)!}$$
$$= \frac{(n+1)!}{3!\,(n-2)!} - \frac{n!\,(n-2)}{3!\,(n-2)!}$$
$$= \frac{(n+1)! - n!\,(n-2)}{3!\,(n-2)!}$$
$$= \frac{n!\,[(n+1)-(n-2)]}{3!\,(n-2)!}$$
$$= \frac{n! \times 3}{3!\,(n-2)!} = \frac{n!}{2!\,(n-2)!}$$
$$= \binom{n}{2}$$

6.
$$\mathbf{v} \times \mathbf{w} = \begin{vmatrix} \mathbf{i} & \mathbf{j} & \mathbf{k} \\ 3 & 2 & -1 \\ -1 & 1 & 4 \end{vmatrix}$$
$$= \mathbf{i}\begin{vmatrix} 2 & -1 \\ 1 & 4 \end{vmatrix} - \mathbf{j}\begin{vmatrix} 3 & -1 \\ -1 & 4 \end{vmatrix} + \mathbf{k}\begin{vmatrix} 3 & 2 \\ -1 & 1 \end{vmatrix}$$
$$= 9\mathbf{i} - 11\mathbf{j} + 5\mathbf{k}$$
$$\mathbf{u}.(\mathbf{v} \times \mathbf{w}) = (-2\mathbf{i} + 0\mathbf{j} + 5\mathbf{k}).(9\mathbf{i} - 11\mathbf{j} + 5\mathbf{k})$$
$$= -18 + 0 + 25$$
$$= 7.$$

7.
$$\int_1^2 \frac{3x+5}{(x+1)(x+2)(x+3)}\,dx$$
$$\frac{3x+5}{(x+1)(x+2)(x+3)} = \frac{A}{x+1} + \frac{B}{x+2} + \frac{C}{x+3}$$
$$3x + 5 = A(x+2)(x+3) + B(x+1)(x+3) + C(x+1)(x+2)$$

$$x = -1 \Rightarrow 2 = 2A \Rightarrow A = 1$$
$$x = -2 \Rightarrow -1 = -B \Rightarrow B = 1$$
$$x = -3 \Rightarrow -4 = 2C \Rightarrow C = -2$$

Hence
$$\frac{3x+5}{(x+1)(x+2)(x+3)} = \frac{1}{x+1} + \frac{1}{x+2} - \frac{2}{x+3}$$
$$\int_1^2 \frac{3x+5}{(x+1)(x+2)(x+3)}\,dx = \int_1^2 \left(\frac{1}{x+1} + \frac{1}{x+2} - \frac{2}{x+3}\right)dx$$
$$= [\ln(x+1) + \ln(x+2) - 2\ln(x+3)]_1^2$$
$$= \ln 3 + \ln 4 - 2\ln 5 - \ln 2 - \ln 3 + 2\ln 4$$
$$= \ln\frac{3 \times 4 \times 4^2}{5^2 \times 2 \times 3} = \ln\frac{32}{25}$$

8. (a) Write the odd integers as: $2n+1$ and $2m+1$ where n and m are integers.
Then
$$(2n+1)(2m+1) = 4nm + 2n + 2m + 1$$
$$= 2(2nm + n + m) + 1$$
which is odd.

(b) Let $n = 1$, $p^1 = p$ which is given as odd.
Assume p^k is odd and consider p^{k+1}.
$$p^{k+1} = p^k \times p$$
Since p^k is assumed to be odd and p is odd, p^{k+1} is the product of two odd integers is therefore odd.
Thus p^{n+1} is an odd integer for all n if p is an odd integer.

9. Let $f(x) = (1 + \sin^2 x)$. Then
$$f(0) = 1$$
$$f'(x) = 2\sin x \cos x \Rightarrow f'(0) = 0$$
$$= \sin 2x$$
$$f''(x) = 2\cos 2x \Rightarrow f''(0) = 2$$
$$f'''(x) = -4\sin 2x \Rightarrow f'''(0) = 0$$
$$f''''(x) = -8\cos 2x \Rightarrow f''''(0) = -8$$

$$f(x) = 1 + 2\frac{x^2}{2!} - 8\frac{x^4}{4!} + \dots$$
$$= 1 + x^2 - \frac{1}{3}x^4 + \dots$$

Alternative 1
$$f(0) = 1$$
$$f'(x) = 2\sin x \cos x \Rightarrow f'(0) = 0$$
$$f''(x) = 2\cos^2 x - 2\sin^2 x \Rightarrow f''(0) = 2$$
$$f'''(x) = 4(-\sin x)\cos x \Rightarrow f'''(0) = 0$$
$$-4\cos x \sin x$$
$$f''''(x) = -8\cos^2 x + 8\sin^2 x \Rightarrow f''''(0) = -8$$
etc

Alternative 2
$$f(x) = (1 + \sin^2 x)$$
$$= 1 + \tfrac{1}{2} - \tfrac{1}{2}\cos 2x$$
$$= \tfrac{1}{2}(3 - \cos 2x)$$
$$= \tfrac{1}{2}\left(3 - \left(1 - \frac{(2x)^2}{2!} + \frac{(2x)^4}{4!} - \dots\right)\right)$$
$$= \tfrac{1}{2}\left(3 - 1 + 2x^2 - \tfrac{2}{3}x^4 - \dots\right)$$
$$= 1 + x^2 - \tfrac{1}{3}x^4 - \dots$$

10. The graph is not symmetrical about the y-axis (or $f(x) \neq f(-x)$) so it is not an even function.
The graph does not have half-turn rotational symmetry (or $f(x) \neq -f(-x)$) so it is not an odd function.
The function is neither even nor odd.

11.
$$\frac{d^2y}{dx^2} + 4\frac{dy}{dx} + 5y = 0$$
$$m^2 + 4m + 5 = 0$$
$$(m+2)^2 = -1$$
$$m = -2 \pm i$$

The general solution is
$$y = e^{-2x}(A\cos x + B\sin x)$$

$$x = 0, y = 3 \qquad 3 = A$$
$$x = \tfrac{\pi}{2}, y = e^{-\pi} \Rightarrow e^{-\pi} = e^{-\pi}\left(3\cos\tfrac{\pi}{2} + B\sin\tfrac{\pi}{2}\right)$$
$$\Rightarrow B = 1$$

The particular solution is:
$$y = e^{-2x}(3\cos x + \sin x).$$

12. Assume $2 + x$ is rational
and let $2 + x = \dfrac{p}{q}$ where p, q are integers.

So
$$x = \frac{p}{q} - 2$$
$$= \frac{p - 2q}{q}$$

Since $p - 2q$ and q are integers, it follows that x is rational. This is a contradiction.

13.
$$y = t^3 - \tfrac{5}{2}t^2 \;\Rightarrow\; \tfrac{dy}{dt} = 3t^2 - 5t$$
$$x = \sqrt{t} = t^{1/2} \;\Rightarrow\; \tfrac{dx}{dt} = \tfrac{1}{2}t^{-1/2}$$
$$\Rightarrow \;\frac{dy}{dx} \;=\; \frac{3t^2 - 5t}{\tfrac{1}{2}t^{-1/2}}$$
$$= 6t^{5/2} - 10t^{3/2}$$
$$\frac{d^2y}{dx^2} = \frac{\frac{d}{dt}\left(\frac{dy}{dx}\right)}{\frac{dx}{dt}}$$
$$= \frac{6 \times \tfrac{5}{2}t^{3/2} - 10 \times \tfrac{3}{2}t^{1/2}}{\tfrac{1}{2}t^{-1/2}}$$
$$= 30t^2 - 30t$$
i.e. $a = 30,\; b = -30$

At a point of inflexion, $\frac{d^2y}{dx^2} = 0 \Rightarrow t = 0$ or 1

But $t > 0 \Rightarrow t = 1 \Rightarrow \frac{dy}{dx} = -4$

and the point of contact is $\left(1, -\tfrac{3}{2}\right)$

Hence the tangent is
$$y + \tfrac{3}{2} = -4(x - 1)$$
i.e. $2y + 8x = 5$

14.
$$\begin{array}{ccc|c} 1 & -1 & 1 & 1 \\ 1 & 1 & 2 & 0 \\ 2 & -1 & a & 2 \end{array}$$
$$\begin{array}{ccc|c} 1 & -1 & 1 & 1 \\ 0 & 2 & 1 & -1 \\ 0 & 1 & a-2 & 0 \end{array}$$
$$\begin{array}{ccc|c} 1 & -1 & 1 & 1 \\ 0 & 2 & 1 & -1 \\ 0 & 0 & 2a-5 & 1 \end{array}$$
$$z = \frac{1}{2a - 5};$$
$$2y + \frac{1}{2a-5} = -1 \;\Rightarrow\; 2y = \frac{-2a+5-1}{2a-5}$$
$$\Rightarrow\; y = \frac{2-a}{2a-5};$$
$$x - \frac{2-a}{2a-5} + \frac{1}{2a-5} = 1$$
$$\Rightarrow\; x = \frac{2a-5}{2a-5} + \frac{1-a}{2a-5} = \frac{a-4}{2a-5}.$$

which exist when $2a - 5 \neq 0$.

From the third row of the final tableau, when $a = 2{\cdot}5$, there are no solutions

When $a = 3$, $x = -1$, $y = -1$, $z = 1$.

$$AB = \begin{pmatrix} 5 & 2 & -3 \\ 1 & 1 & -1 \\ -3 & -1 & 2 \end{pmatrix}\begin{pmatrix} 1 \\ 0 \\ 2 \end{pmatrix} = \begin{pmatrix} -1 \\ -1 \\ 1 \end{pmatrix}$$

From above, we have $C\begin{pmatrix} -1 \\ -1 \\ 1 \end{pmatrix} = \begin{pmatrix} 1 \\ 0 \\ 2 \end{pmatrix}$ and

also $A\begin{pmatrix} 1 \\ 0 \\ 2 \end{pmatrix} = \begin{pmatrix} -1 \\ -1 \\ 1 \end{pmatrix}$ which suggests $AC = I$ and

this can be verified directly. Hence
A is the inverse of C (or vice versa).

15.
$$(x^2)^2 = 8x \Rightarrow x^4 = 8x \Rightarrow x = 0,\, 2$$
$$\text{Area} = 4\int_0^2 (\sqrt{8x} - x^2)\, dx$$
$$= 4\left[\sqrt{8}\left(\tfrac{2}{3}x^{3/2}\right) - \tfrac{1}{3}x^3\right]_0^2$$
$$= 4\left[\frac{16}{3} - \frac{8}{3}\right] = \frac{32}{3}$$

Volume of revolution about the y-axis $= \pi\!\int x^2\, dy$.
So in this case, we need to calculate
two volumes and subtract:
$$V = \pi\left[\int_0^4 y\, dy\right] - \pi\left[\int_0^4 \tfrac{y^4}{64}\, dy\right]$$
$$= \pi\left[\frac{y^2}{2}\right]_0^4 - \pi\left[\frac{y^5}{320}\right]_0^4$$
$$= \pi\left[8 - \frac{64 \times 4^2}{320}\right]$$
$$= \frac{40 - 16}{5}\pi$$
$$= \frac{24\pi}{5}\; (\approx 15)$$

16.
$$z^3 = r^3(\cos 3\theta + i\sin 3\theta)$$
$$\left(\cos\tfrac{2\pi}{3} + i\sin\tfrac{2\pi}{3}\right)^3 = \cos 2\pi + i\sin 2\pi$$
$$a = 1;\; b = 0$$
Method 1
$$r^3(\cos 3\theta + i\sin 3\theta) = 8$$
$$r^3\cos 3\theta = 8 \text{ and } r^3\sin 3\theta = 0$$
$$\Rightarrow r = 2;\; 3\theta = 0,\, 2\pi,\, 4\pi$$
Roots are 2, $2\left(\cos\tfrac{2\pi}{3} + i\sin\tfrac{2\pi}{3}\right)$, $2\left(\cos\tfrac{4\pi}{3} + i\sin\tfrac{4\pi}{3}\right)$.
In cartesian form: 2, $\left(-1 + i\sqrt{3}\right)$, $\left(-1 - i\sqrt{3}\right)$
Method 2
$$z^3 - 8 = 0$$
$$(z - 2)(z^2 + 2z + 4) = 0$$
$$(z - 2)\left((z + 1)^2 + (\sqrt{3})^2\right) = 0$$
so the roots are: 2, $\left(-1 + i\sqrt{3}\right)$, $\left(-1 - i\sqrt{3}\right)$

(a) $\qquad z_1 + z_2 + z_3 = 0$

(b) Since $z_1^3 = z_2^3 = z_3^3 = 8$
it follows that
$$z_1^6 + z_2^6 + z_3^6 = (z_1^3)^2 + (z_2^3)^2 + (z_3^3)^2$$
$$= 3 \times 64 = 192$$

ADVANCED HIGHER MATHEMATICS 2011

1.
$$\frac{13 - x}{x^2 + 4x - 5} = \frac{13 - x}{(x - 1)(x + 5)}$$
$$= \frac{A}{x - 1} + \frac{B}{x + 5}$$
$$13 - x = A(x + 5) + B(x - 1)$$
$$x = 1 \Rightarrow 12 = 6A \Rightarrow A = 2$$
$$x = -5 \Rightarrow 18 = -6B \Rightarrow B = -3$$
Hence $\dfrac{13 - x}{x^2 + 4x - 5} = \dfrac{2}{x - 1} - \dfrac{3}{x + 5}$
$$\int \frac{13 - x}{x^2 + 4x - 5}dx = \int \frac{2}{x - 1}dx - \int \frac{3}{x + 5}dx$$
$$= 2\ln|x - 1| - 3\ln|x + 5| + c$$

2. $\left(\frac{1}{2}x - 3\right)^4 = {}^4C_0\left(\frac{x}{2}\right)^4 + {}^4C_1\left(\frac{x}{2}\right)^3(-3) + $
$\left. {}^4C_2\left(\frac{x}{2}\right)^2(-3)^2 + {}^4C_3\left(\frac{x}{2}\right)(-3)^3 + {}^4C_4(-3)^4\right\}$
$$= \left(\frac{x}{2}\right)^4 + 4\left(\frac{x}{2}\right)^3(-3) + 6\left(\frac{x}{2}\right)^2(-3)^2 + 4\left(\frac{x}{2}\right)(-3)^3 + (-3)^4$$
$$= \frac{x^4}{16} - \frac{3x^3}{2} + \frac{27x^2}{2} - 54x + 81.$$

3. (a) *Method 1*
$$y + e^y = x^2$$
$$\frac{dy}{dx} + e^y\frac{dy}{dx} = 2x$$
$$\frac{dy}{dx}(1 + e^y) = 2x \Rightarrow \frac{dy}{dx} = \frac{2x}{(1 + e^y)}$$

Method 2
$$\ln(y + e^y) = 2\ln x$$
$$\frac{(1 + e^y)\frac{dy}{dx}}{y + e^y} = \frac{2}{x}$$
$$\frac{dy}{dx} = \frac{2(y + e^y)}{x(1 + e^y)}$$

Method 3
$$y + e^y = x^2 \Rightarrow e^y = x^2 - y \Rightarrow y = \ln(x^2 - y)$$
$$\frac{dy}{dx} = \frac{2x - \frac{dy}{dx}}{x^2 - y}$$
$$\frac{dy}{dx}(x^2 - y) = 2x - \frac{dy}{dx} \Rightarrow \frac{dy}{dx}(x^2 - y + 1) = 2x$$
$$\frac{dy}{dx} = \frac{2x}{x^2 - y + 1}$$

(b) *Method 1*
$$f(x) = \sin x \cos^3 x$$
$$f'(x) = \cos^4 x + \sin x(-3\cos^2 x \sin x)$$
$$= \cos^4 x - 3\cos^2 x \sin^2 x$$

Method 2
$$f(x) = \sin x \cos^3 x$$
$$\ln(f(x)) = \ln\sin x + \ln(\cos^3 x)$$
$$\frac{f'(x)}{f(x)} = \frac{\cos x}{\sin x} - \frac{3\cos^2 x \sin x}{\cos^3 x}$$
$$= \frac{\cos x}{\sin x} - \frac{3\sin x}{\cos x}$$
$$f'(x) = \left(\frac{\cos x}{\sin x} - \frac{3\sin x}{\cos x}\right)\sin x \cos^3 x$$
$$= \cos^4 x - 3\sin^2 x \cos^2 x$$

4. (a) Singular when the determinant is 0.
$$1\det\begin{pmatrix} 0 & 2 \\ \lambda & 6 \end{pmatrix} - 2\det\begin{pmatrix} 3 & 2 \\ -1 & 6 \end{pmatrix} + (-1)\det\begin{pmatrix} 3 & 0 \\ -1 & \lambda \end{pmatrix} = 0$$
$$-2\lambda - 2(18 + 2) - 1(3\lambda - 0) = 0$$
$$-5\lambda - 40 = 0 \text{ when } \lambda = -8$$

(b) From A, $A' = \begin{pmatrix} 2 & 3\alpha + 2\beta & -1 \\ 2\alpha - \beta & 4 & 3 \\ -1 & 3 & 2 \end{pmatrix}$.

Hence $2\alpha - \beta = -1$ and $3\alpha + 2\beta = -5$.
$$4\alpha - 2\beta = -2$$
$$3\alpha + 2\beta = -5$$
$$7\alpha = -7 \Rightarrow \alpha = -1 \text{ and } \beta = -1.$$

5. Let $f(x) = (1 + x)^{\frac{1}{2}}$, then
$$f(x) = (1 + x)^{\frac{1}{2}} \Rightarrow f(0) = 1$$
$$f'(x) = \frac{1}{2}(1 + x)^{-\frac{1}{2}} \Rightarrow f'(0) = \frac{1}{2}$$
$$f''(x) = -\frac{1}{4}(1 + x)^{-\frac{3}{2}} \Rightarrow f''(0) = -\frac{1}{4}$$
$$f'''(x) = \frac{3}{8}(1 + x)^{-\frac{5}{2}} \Rightarrow f'''(0) = \frac{3}{8}$$
Hence
$$\sqrt{1 + x} = 1 + \frac{1}{2}x - \frac{1}{4} \times \frac{x^2}{2} + \frac{3}{8} \times \frac{x^3}{6} - \dots$$
$$= 1 + \frac{1}{2}x - \frac{x^2}{8} + \frac{x^3}{16} - \dots$$
and replacing x by x^2 gives
$$\sqrt{1 + x^2} = 1 + \frac{1}{2}x^2 - \frac{x^4}{8} + \frac{x^6}{16} - \dots$$
Thus
$$\sqrt{(1 + x)(1 + x^2)} =$$
$$\left(1 + \frac{1}{2}x - \frac{x^2}{8} + \frac{x^3}{16} - \dots\right)\left(1 + \frac{1}{2}x^2 - \frac{x^4}{8} + \frac{x^6}{16} - \dots\right)$$
$$= 1 + \frac{1}{2}x + \frac{1}{2}x^2 - \frac{1}{8}x^2 + \frac{1}{4}x^3 + \frac{1}{16}x^3 + \dots$$
$$= 1 + \frac{1}{2}x + \frac{3}{8}x^2 + \frac{5}{16}x^3 + \dots$$

6. Reflect in the line $y = x$ to get

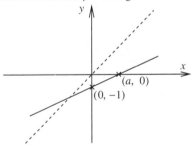

Now apply the modulus function

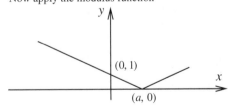

7. *Method 1*

$$y = \frac{e^{\sin x}(2 + x)^3}{\sqrt{1 - x}}$$

$$\Rightarrow \ln y = \ln\left(e^{\sin x}(2 + x)^3\right) - \ln\left(\sqrt{1 - x}\right)$$

$$= \sin x + 3 \ln(2 + x) - \tfrac{1}{2} \ln(1 - x)$$

$$\Rightarrow \frac{1}{y}\frac{dy}{dx} = \cos x + \frac{3}{2 + x} + \frac{1}{2(1 - x)}$$

$$\frac{dy}{dx} = y\left(\cos x + \frac{3}{2 + x} + \frac{1}{2(1 - x)}\right)$$

When $x = 0$, $y = 8 \Rightarrow$

gradient $= 8\left(1 + \frac{3}{2} + \frac{1}{2}\right) = 24$.

Method 2

$$y = \frac{e^{\sin x}(2 + x)^3}{\sqrt{1 - x}} \Rightarrow$$

$$\frac{dy}{dx} = \frac{\frac{d}{dx}\left(e^{\sin x}(2+x)^3\right)\sqrt{1-x} - e^{\sin x}(2+x)^3\left(-\frac{1}{2}\frac{1}{\sqrt{1-x}}\right)}{(1-x)}$$

$$= \frac{\left[\cos x\, e^{\sin x}(2+x)^3 + 3e^{\sin x}(2+x)^2\right](1-x)}{(1-x)^{3/2}} + \frac{e^{\sin x}(2+x)^3}{2(1-x)^{3/2}}$$

When $x = 0$, gradient $= \frac{(2^3 + 3 \times 2^2)}{1} + \frac{2^3}{2} = 20 + 4 = 24$

Method 3

$$y = \frac{e^{\sin x}(2+x)^3}{\sqrt{1-x}}$$

$$y\sqrt{1-x} = e^{\sin x}(2 + x)^3$$

$$\sqrt{1-x}\frac{dy}{dx} - \tfrac{1}{2}y(1-x)^{-1/2} = \cos x\, e^{\sin x}(2+x)^3 + 3e^{\sin x}(2+x)^2$$

when $x = 0$, $y = \frac{e^0 2^3}{1} = 8$. This leads to $\frac{dy}{dx} = 24$

8.

$$\sum_{r=1}^{n} r^3 - \left(\sum_{r=1}^{n} r\right)^2 = \frac{n^2(n+1)^2}{4} - \left(\frac{n(n+1)}{2}\right)^2 = 0$$

$$\sum_{r=1}^{n} r^3 + \left(\sum_{r=1}^{n} r\right)^2 = \frac{n^2(n+1)^2}{4} + \left(\frac{n(n+1)}{2}\right)^2$$

$$= \frac{n^2(n+1)^2}{4} + \frac{n^2(n+1)^2}{4}$$

$$= \frac{n^2(n+1)^2}{2}$$

9. *Method 1*

$$\frac{dy}{dx} = 3(1 + y)\sqrt{1 + x}$$

$$\int \frac{dy}{1 + y} = 3\int(1 + x)^{\frac{1}{2}} dx$$

$$\ln(1 + y) = 2(1 + x)^{\frac{3}{2}} + c$$

$$1 + y = \exp\left(2(1 + x)^{\frac{3}{2}} + c\right)$$

$$y = \exp\left(2(1 + x)^{\frac{3}{2}} + c\right) - 1.$$

$$= A \exp\left(2(1 + x)^{\frac{3}{2}}\right) - 1.$$

Method 2

$$\frac{dy}{dx} - 3\sqrt{1 + x}\, y = 3\sqrt{1 + x}$$

Integrating Factor

$$\exp\left(-3\int\sqrt{1 + x}\, dx\right) = \exp\left(-2(1 + x)^{3/2}\right)$$

$$\frac{d}{dx}\left(y\exp\left(-2(1 + x)^{3/2}\right)\right) = 3\sqrt{1 + x}\left(\exp\left(-2(1 + x)^{3/2}\right)\right)$$

$$y\left(\exp\left(-2(1 + x)^{3/2}\right)\right) = -\int 3\sqrt{1 + x}\,\exp\left(-2(1 + x)^{3/2}\right) dx$$

$$= -\exp\left(-2(1 + x)^{3/2}\right) + c$$

$$y = -1 + c\exp\left(2(1 + x)^{3/2}\right)$$

10. Let $z = x + iy$ so $z - 1 = (x - 1) + iy$.

$|z - 1|^2 = (x - 1)^2 + y^2 = 9$.

The locus is the circle with centre $(1, 0)$ and radius 3.

11. (a) $\int_0^{\pi/4}(\sec x - x)(\sec x + x)\,dx = \int_0^{\pi/4}(\sec^2 x - x^2)\,dx$

$$= \left[\tan x - \frac{x^3}{3}\right]_0^{\frac{\pi}{4}}$$

$$= \left[1 - \frac{1}{3}\frac{\pi^3}{64}\right] - [0]$$

$$= 1 - \frac{\pi^3}{192}.$$

(b) *Method 1*

Let $u = 7x^2$, then $du = 14x\,dx$.

$$\int \frac{x}{\sqrt{1 - 49x^4}}\, dx = \frac{1}{14}\int \frac{du}{\sqrt{1 - u^2}}$$

$$= \frac{1}{14}\sin^{-1}u + c$$

$$= \frac{1}{14}\sin^{-1}7x^2 + c$$

Method 2

$$\int \frac{x}{\sqrt{1 - 49x^4}}\, dx = \frac{1}{14}\int \frac{14x\,dx}{\sqrt{1 - (7x^2)^2}}$$

$$= \frac{1}{14}\sin^{-1}7x^2 + c$$

12. For $n = 2$, $8^2 + 3^0 = 64 + 1 = 65$.

True when $n = 2$.

Assume true for k, i.e. that $8^k + 3^{k-2}$ is divisible by 5, i.e. can be expressed as $5p$ for an integer p.

Now consider $8^{k+1} + 3^{k-1}$

$$= 8 \times 8^k + 3^{k-1}$$

$$= 8 \times \left(5p - 3^{k-2}\right) + 3^{k-1}$$

$$= 40p - 3^{k-2}(8 - 3)$$

$$= 5\left(8p - 3^{k-2}\right) \text{ which is divisible by 5.}$$

So, since it is true for $n = 2$, it is true for all $n \geqslant 2$.

13. *Method 1*

Let d be the common difference. Then

$$u_3 = 1 = a + 2d \qquad \text{and} \qquad u_2 = \frac{1}{a} = a + d$$

$$1 = a + 2\left(\frac{1}{a} - a\right)$$

$$a = a^2 + 2 - 2a^2$$

$$a^2 + a - 2 = 0$$

$$(a + 2)(a - 1) = 0 \Rightarrow a = -2 \text{ since } a < 0.$$

$a = -2$ gives $2d = 3$ and hence $d = \frac{3}{2}$.

Method 2

$$u_1 = a,\ u_2 = \frac{1}{a},\ u_3 = 1$$

$$\Rightarrow \frac{1}{a} - a = 1 - \frac{1}{a}$$

$$\Rightarrow 1 - a^2 = a - 1$$

$$\Rightarrow a^2 + a - 2 = 0$$

$$(a + 2)(a - 1) = 0 \Rightarrow a = -2 \text{ since } a < 0.$$

$$d = u_3 - u_2 = 1 - \frac{1}{a} = \frac{3}{2}$$

13. *(continued)*

$$S_n = \frac{n}{2}[2a + (n-1)d]$$
$$= \frac{n}{2}\left[-4 + \frac{3}{2}n - \frac{3}{2}\right]$$
$$= \frac{1}{4}[3n^2 - 11n]$$
$$\therefore 3n^2 - 11n > 4000$$
$$n^2 - \frac{11}{3}n > \frac{4000}{3}$$
$$\left(n - \frac{11}{6}\right)^2 > \frac{48000}{36} + \frac{121}{36} = \frac{48121}{36}$$
$$n - \frac{11}{6} > \frac{\sqrt{48121}}{6}$$
$$n > \frac{\sqrt{48121} + 11}{6} \approx 38.39$$

So the least value of n is 39.

14. Auxiliary equation
$$m^2 - m - 2 = 0$$
$$(m-2)(m+1) = 0$$
$$m = -1 \text{ or } 2$$
Complementary function is: $y = Ae^{-x} + Be^{2x}$

The particular integral has the form $y = Ce^x + D$
$$y = Ce^x + D \Rightarrow \frac{dy}{dx} = Ce^x$$
$$\Rightarrow \frac{d^2y}{dx^2} = Ce^x$$
Hence we need:
$$\frac{d^2y}{dx^2} - \frac{dy}{dx} - 2y = e^x + 12$$
$$[Ce^x] - [Ce^x] - 2[Ce^x + D] = e^x + 12$$
$$-2Ce^x - 2D = e^x + 12$$
Hence $C = -\frac{1}{2}$ and $D = -6$.
So the General Solution is
$$y = Ae^{-x} + Be^{2x} - \frac{1}{2}e^x - 6.$$
$x = 0$ and $y = -\frac{3}{2} \Rightarrow A + B - \frac{1}{2} - 6 = -\frac{3}{2}$
$x = 0$ and $\frac{dy}{dx} = \frac{1}{2} \Rightarrow -A + 2B - \frac{1}{2} = \frac{1}{2}$
$$3B - 7 = -1 \Rightarrow B = 2 \Rightarrow A = 3$$
So the particular solution is
$$y = 3e^{-x} + 2e^{2x} - \frac{1}{2}e^x - 6.$$

15. *(a)* In terms of a parameter s, L_1 is given by
$$x = 1 + ks, \ y = -s, \ z = -3 + s$$

In terms of a parameter t, L_2 is given by
$$x = 4 + t, \ y = -3 + t, \ z = -3 + 2t$$

Equating the y coordinates and equating the z coordinates:

$$-s = -3 + t$$
$$-3 + s = -3 + 2t$$
Adding these
$$-3 = -6 + 3t \Rightarrow t = 1 \Rightarrow s = 2.$$

From the x coordinates
$$1 + ks = 4 + t$$
Using the values of s and t
$$1 + 2k = 5 \Rightarrow k = 2$$

The point of intersection is: $(5, -2, -1)$.

(b) L_1 has direction $2\mathbf{i} - \mathbf{j} + \mathbf{k}$.
L_2 has direction $\mathbf{i} + \mathbf{j} + 2\mathbf{k}$.

Let the angle between L_1 and L_2 be θ, then
$$\cos\theta = \frac{(2\mathbf{i} - \mathbf{j} + \mathbf{k}) \cdot (\mathbf{i} + \mathbf{j} + 2\mathbf{k})}{|2\mathbf{i} - \mathbf{j} + \mathbf{k}||\mathbf{i} + \mathbf{j} + 2\mathbf{k}|}$$
$$= \frac{2 - 1 + 2}{\sqrt{6}\sqrt{6}} = \frac{3}{6} = \frac{1}{2}$$
$$\theta = 60°$$
The angle between L_1 and L_2 is $60°$.

16. *(a)* $I_n = \int_0^1 \frac{1}{(1+x^2)^n} dx$
$$= \int_0^1 1 \times (1+x^2)^{-n} dx$$
$$= \left[(1+x^2)^{-n} \int 1\,dx\right]_0^1 + \int_0^1 \left(2nx(1+x^2)^{-n-1} \int 1\,dx\right) dx$$
$$= \left[x(1+x^2)^{-n}\right]_0^1 + \int_0^1 2nx^2(1+x^2)^{-n-1} dx$$
$$= \frac{1}{2^n} - 0 + 2n\int_0^1 x^2(1+x^2)^{-n-1} dx$$
$$= \frac{1}{2^n} + 2n\int_0^1 \frac{x^2}{(1+x^2)^{n+1}} dx.$$

(b) $\dfrac{A}{(1+x^2)^n} + \dfrac{B}{(1+x^2)^{n+1}} = \dfrac{x^2}{(1+x^2)^{n+1}}$
$$\Rightarrow A(1+x^2) + B = x^2$$
$$\Rightarrow A = 1, B = -1$$
$$\frac{1}{(1+x^2)^n} + \frac{-1}{(1+x^2)^{n+1}} = \frac{x^2}{(1+x^2)^{n+1}}$$
$$I_n = \frac{1}{2^n} + 2n\int_0^1 \frac{x^2}{(1+x^2)^{n+1}} dx.$$
$$= \frac{1}{2^n} + 2n\int_0^1 \frac{1}{(1+x^2)^n} dx + 2n\int_0^1 \frac{-1}{(1+x^2)^{n+1}} dx$$
$$= \frac{1}{2^n} + 2nI_n - 2nI_{n+1}$$
$$2nI_{n+1} = \frac{1}{2^n} + (2n - 1)I_n$$
$$I_{n+1} = \frac{1}{n \times 2^{n+1}} + \left(\frac{2n-1}{2n}\right)I_n.$$

(c) $\displaystyle\int_0^1 \frac{1}{(1+x^2)^3} dx = I_3$
$$= \frac{1}{16} + \frac{3}{4}I_2$$
$$= \frac{1}{16} + \frac{3}{4}\left(\frac{1}{4} + \frac{1}{2}I_1\right)$$
$$= \frac{1}{4} + \frac{3}{8}\int_0^1 \frac{1}{1+x^2} dx$$
$$= \frac{1}{4} + \frac{3}{8}\left[\tan^{-1}x\right]_0^1$$
$$= \frac{1}{4} + \frac{3}{8}\frac{\pi}{4} = \frac{1}{4} + \frac{3\pi}{32}.$$

ADVANCED HIGHER MATHEMATICS 2012

1. (a) $f(x) = \dfrac{3x + 1}{x^2 + 1}$

$$f'(x) = \frac{3(x^2 + 1) - (3x + 1)2x}{(x^2 + 1)^2}$$

$$= \frac{3x^2 + 3 - 6x^2 - 2x}{(x^2 + 1)^2}$$

$$= \frac{-3x^2 - 2x + 3}{(x^2 + 1)^2}$$

(b) $g(x) = \cos^2 x \, e^{\tan x}$

$$g'(x) = 2\cos x(-\sin x)e^{\tan x} + (\cos^2 x)(\sec^2 x)e^{\tan x}$$

$$= -\sin 2x \, e^{\tan x} + e^{\tan x}$$

$$= (1 - \sin 2x)\, e^{\tan x}$$

2. $a = 2048$ and $ar^3 = 256$

$$\Rightarrow r^3 = \tfrac{1}{8}$$
$$\Rightarrow r = \tfrac{1}{2}.$$

$$S_n = \frac{a(1 - r^n)}{1 - r}$$

$$\Rightarrow \frac{1 - (\tfrac{1}{2})^n}{1 - \tfrac{1}{2}} = \frac{4088}{2048}$$

$$= \frac{511}{256}$$

$$\Rightarrow 1 - \left(\frac{1}{2}\right)^n = \frac{511}{256} \times \frac{1}{2} = \frac{511}{512}$$

$$\frac{1}{2^n} = 1 - \frac{511}{512} = \frac{1}{512}$$

$$\Rightarrow 2^n = 512 \Rightarrow n = 9$$

3. Since w is a root, $\bar{w} = -1 - 2i$ is also a root.

The corresponding factors are

$$(z + 1 - 2i) \text{ and } (z + 1 + 2i)$$

from which

$$\big((z + 1) - 2i\big)\big((z + 1) + 2i\big) = (z + 1)^2 + 4$$
$$= z^2 + 2z + 5$$
$$z^3 + 5z^2 + 11z + 15 = (z^2 + 2z + 5)(z + 3)$$

The roots are $(-1 + 2i)$, $(-1 - 2i)$ and -3.

4. The general term is given by:

$$\binom{9}{r}(2x)^{9-r}\left(-\frac{1}{x^2}\right)^r$$

$$= \binom{9}{r} \times \frac{2^{9-r}x^{9-r}(-1)^r}{x^{2r}}$$

$$= \binom{9}{r} \times (-1)^r 2^{9-r} x^{9-3r}$$

The term independent of x occurs when

$$9 - 3r = 0, \text{ i.e. when } r = 3.$$

The term is: $\dfrac{9!}{6!\,3!}(-1)^3 2^6$

$$= -5376.$$

5. *Method 1*

$$\overrightarrow{PQ} = 3\mathbf{i} + \mathbf{j} + 4\mathbf{k} \text{ and } \overrightarrow{QR} = 2\mathbf{i} - 2\mathbf{j} - 2\mathbf{k}$$

A normal to the plane:

$$\overrightarrow{PQ} \times \overrightarrow{QR} = \begin{vmatrix} \mathbf{i} & \mathbf{j} & \mathbf{k} \\ 3 & 1 & 4 \\ 2 & -2 & -2 \end{vmatrix}$$

$$= \mathbf{i}\begin{vmatrix} 1 & 4 \\ -2 & -2 \end{vmatrix} - \mathbf{j}\begin{vmatrix} 3 & 4 \\ 2 & -2 \end{vmatrix} + \mathbf{k}\begin{vmatrix} 3 & 1 \\ 2 & -2 \end{vmatrix}$$

$$= 6\mathbf{i} + 14\mathbf{j} - 8\mathbf{k}$$

Hence the equation has the form:

$$6x + 14y - 8z = d.$$

The plane passes through $P(-2, 1, -1)$ so

$$d = -12 + 14 + 8 = 10$$

which gives an equation $6x + 14y - 8z = 10$

i.e. $3x + 7y - 4z = 5$.

Method 2

A plane has an equation of the form $ax + by + cz = d$.

Using the points P, Q, R we get

$$-2a + b - c = d$$
$$a + 2b + 3c = d$$
$$3a + c = d$$

Using Gaussian elimination to solve these we have

$$\begin{vmatrix} -2 & 1 & -1 & d \\ 1 & 2 & 3 & d \\ 3 & 0 & 1 & d \end{vmatrix} \Rightarrow \begin{vmatrix} -2 & 1 & -1 & d \\ 0 & 5 & 5 & 3d \\ 0 & 6 & 8 & 2d \end{vmatrix}$$

$$\Rightarrow \begin{vmatrix} -2 & 1 & -1 & d \\ 0 & 5 & 5 & 3d \\ 0 & 0 & 2 & -\tfrac{8}{5}d \end{vmatrix}$$

$$\Rightarrow c = -\frac{4}{5}d, \quad b = \frac{7}{5}d, \quad a = \frac{3}{5}d$$

These give the equation

$$\left(\tfrac{3}{5}d\right)x + \left(\tfrac{7}{5}d\right)y + \left(-\tfrac{4}{5}d\right)z = d$$

i.e. $3x + 7y - 4z = 5$

6. *Method 1*

A $e^x = 1 + x + \tfrac{x^2}{2} + \tfrac{x^3}{6} + \dots$

B $(1 + e^x)^2 = 1 + 2e^x + e^{2x}$

C $= 1 + 2\left(1 + x + \tfrac{x^2}{2} + \tfrac{x^3}{6} + \dots\right) + \left(1 + 2x + \tfrac{(2x)^2}{2} + \tfrac{(2x)^3}{6} + \dots\right)$

D $= 1 + 2 + 2x + x^2 + \tfrac{1}{3}x^3 + 1 + 2x + 2x^2 + \tfrac{4}{3}x^3 + \dots$

E $= 4 + 4x + 3x^2 + \tfrac{5}{3}x^3 + \dots$

Method 2

$$e^x = 1 + x + \tfrac{x^2}{2} + \tfrac{x^3}{6} + \dots$$

$$(1 + e^x) = 2 + x + \tfrac{x^2}{2} + \tfrac{x^3}{6} + \dots$$

$$(1 + e^x)^2 = \left(2 + x + \tfrac{x^2}{2} + \tfrac{x^3}{6} + \dots\right)\left(2 + x + \tfrac{x^2}{2} + \tfrac{x^3}{6} + \dots\right)$$

$$= 4 + 4x + 3x^2 + \tfrac{1}{3}x^3 + \tfrac{1}{2}x^3 + \tfrac{1}{2}x^3 + \tfrac{1}{3}x^3 + \dots$$

$$= 4 + 4x + 3x^2 + \tfrac{5}{3}x^3 + \dots$$

Method 3

$$e^x = 1 + x + \tfrac{x^2}{2} + \tfrac{x^3}{6} + \dots$$

$$f(x) = (1 + e^x)^2 \qquad f(0) = 4$$

$$f'(x) = 2e^x(1 + e^x) \quad f'(0) = 4$$

$$= 2e^x + 2e^{2x}$$

$$f''(x) = 2e^x + 4e^{2x} \quad f''(0) = 6$$

$$f'''(x) = 2e^x + 8e^{2x} \quad f'''(0) = 10$$

$$f(x) = f(0) + f'(0)x + f''(0)\tfrac{x^2}{2} + f'''(0)\tfrac{x^3}{6} + \dots$$

$$(1 + e^x)^2 = 4 + 4x + 3x^2 + \tfrac{5}{3}x^3 + \dots$$

7. (a)

$$y = |x + 2|$$

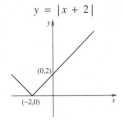

(b)

8. $x = 4 \sin \theta \Rightarrow dx = 4 \cos \theta \, d$

$$\int_0^2 \sqrt{16 - x^2} \, dx$$
$$= \int_0^{\pi/6} \sqrt{16 - (4 \sin \theta)^2} \cdot 4 \cos \theta \, d\theta$$
$$= \int_0^{\pi/6} \sqrt{16(1 - \sin^2 \theta)} \cdot 4 \cos \theta \, d\theta$$
$$= \int_0^{\pi/6} \sqrt{16 \cos^2 \theta} \cdot 4 \cos \theta \, d$$
$$= \int_0^{\pi/6} 16 \cos^2 \theta \, d\theta$$
$$= 8 \int_0^{\pi/6} (1 + \cos 2\theta) \, d\theta$$
$$= 8 \left[\theta + \tfrac{1}{2} \sin 2\theta\right]_0^{\pi/6}$$
$$= \tfrac{8\pi}{6} + 4 \sin \tfrac{\pi}{3}$$
$$= \tfrac{4\pi}{3} + 2\sqrt{3} \ (\approx 7 \cdot 65)$$

9. *Method 1*

$$A + A^{-1} = I$$
$$A^2 + I = A$$

Hence $A^2 + I = I - A^{-1}$

$$A^2 = -A^{-1}$$
$$A^3 = -I, \text{ i.e. } k = -1$$

Method 2

$$A + A^{-1} = I$$
$$A = I - A^{-1}$$
$$A^2 = I - 2A^{-1} + (A^{-1})^2$$
$$A^3 = A - 2I + A^{-1}$$
$$A^3 = (A + A^{-1}) - 2I = I - 2I$$

Hence $A^3 = -I, \text{ i.e. } k = -1$

10. *Method 1*

$$1234 = 7 \times 176 + 2$$
$$176 = 7 \times 25 + 1$$
$$25 = 7 \times 3 + 4$$

Hence

$$1234_{10} = 3412_7$$

Method 2

$$1234 = 7 \times 176 + 2$$
$$= 7 \times (7 \times 25 + 1) + 2$$
$$= 7 \times (7 \times (7 \times 3 + 4) + 1) + 2$$
$$= 3 \times 7^3 + 4 \times 7^2 + 1 \times 7 + 2$$

Hence

$$1234_{10} = 3412_7$$

11. (a) $\dfrac{d}{dx}(\sin^{-1} x) = \dfrac{1}{\sqrt{1 - x^2}}$

(b) $\int \sin^{-1} x \cdot \frac{x}{\sqrt{1-x^2}} \, dx =$

$$\sin^{-1} x \int \frac{x}{\sqrt{1-x^2}} \, dx - \int \left(\frac{d}{dx}(\sin^{-1} x) \int \frac{x}{\sqrt{1-x^2}} \, dx\right) dx$$
$$= \sin^{-1} x \int \frac{x}{\sqrt{1-x^2}} \, dx - \int \left(\frac{1}{\sqrt{1-x^2}} \int \frac{x}{\sqrt{1-x^2}} \, dx\right) dx$$
$$= \sin^{-1} x \left(-\sqrt{1 - x^2}\right) - \int \left(\frac{1}{\sqrt{1-x^2}} \left(-\sqrt{1 - x^2}\right)\right) dx$$
$$= \sin^{-1} x \left(-\sqrt{1 - x^2}\right) - \int (-1) \, dx$$
$$= x - \sin^{-1} x \cdot \sqrt{1 - x^2} + c$$

12. $\dfrac{dr}{dt} = -0.02; \qquad \dfrac{dh}{dt} = 0.01$

$$V = r^2 \pi h$$
$$\frac{dV}{dt} = \pi \left(2r \frac{dr}{dt}\right) h + \pi r^2 \frac{dh}{dt}$$
$$= \pi (2 \times 0.6 \times (-0.02) \times 2 + 0.36 \times 0.01)$$
$$= \pi (-0.048 + 0.0036)$$
$$= -0.0444 \pi (\approx -0.14)$$

The rate of change in the volume is
$-0.0444 \pi \, \text{m}^3 \, \text{s}^{-1}$.

13. $x = 2t + \tfrac{1}{2}t^2 \qquad \Rightarrow \qquad \tfrac{dx}{dt} = 2 + t$

$y = \tfrac{1}{3}t^3 - 3t \qquad \Rightarrow \qquad \tfrac{dy}{dt} = t^2 - 3$

$$\frac{dy}{dx} = \frac{t^2 - 3}{2 + t}$$
$$\frac{d}{dt}\left(\frac{dy}{dx}\right) = \frac{2t(2 + t) - (t^2 - 3)}{(2 + t)^2} = \frac{t^2 + 4t + 3}{(2 + t)^2}$$
$$\frac{d^2y}{dx^2} = \frac{t^2 + 4t + 3}{(2 + t)^2} \times \frac{1}{2 + t} = \frac{t^2 + 4t + 3}{(2 + t)^3}$$

Stationary points when $\frac{dy}{dx} = 0$, i.e.

$$t^2 - 3 = 0 \Rightarrow t = \pm\sqrt{3}$$

When $t = \sqrt{3}$, $\dfrac{d^2y}{dx^2} = \dfrac{3 + 4\sqrt{3} + 3}{(2 + \sqrt{3})^3} > 0$

which gives a minimum.

When $t = -\sqrt{3}$, $\dfrac{d^2y}{dx^2} = \dfrac{3 - 4\sqrt{3} + 3}{(2 - \sqrt{3})^3} < 0$

which gives a maximum.

At a point of inflexion, $\frac{d^2y}{dx^2} = 0$.

In this case, that means

$$t^2 + 4t + 3 = (t + 1)(t + 3) = 0$$

and this has exactly two roots.

Note that this is a slimmed-down version of the complete story of points of inflexion.

14. (a)

$$\begin{vmatrix} 4 & 0 & 6 \\ 2 & -2 & 4 \\ -1 & 1 & \lambda \end{vmatrix}\begin{vmatrix} 1 \\ -1 \\ 2 \end{vmatrix}$$

$$\begin{vmatrix} 4 & 0 & 6 \\ 0 & 4 & -2 \\ 0 & 4 & 6+4\lambda \end{vmatrix}\begin{vmatrix} 1 \\ 3 \\ 9 \end{vmatrix}$$

$$\begin{vmatrix} 4 & 0 & 6 \\ 0 & 4 & -2 \\ 0 & 0 & 8+4\lambda \end{vmatrix}\begin{vmatrix} 1 \\ 3 \\ 6 \end{vmatrix}$$

$$z = \frac{6}{8+4\lambda} = \frac{3}{2(2+\lambda)}$$

$$4y = 3 + 2z \Rightarrow 4y = \frac{18+6\lambda}{4+2\lambda}$$

$$\Rightarrow y = \frac{3\lambda+9}{4(2+\ )\lambda}$$

$$4x = 1 - 6z \Rightarrow 4x = \frac{2\lambda-14}{4+2\lambda}$$

$$\Rightarrow x = \frac{\lambda-7}{4(2+\lambda)}$$

(b) When $\lambda = -2$, the final row gives $0 = 6$ which is inconsistent.
There are no solutions.

(c) $\lambda = -2.1$; $x = 22.75$; $y = -6.75$; $z = -15$
Although the values of λ are close, the values of x, y and z are quite different. The system is **ill-conditioned** near $\lambda = -2$.

15. (a)

$$\frac{1}{(x-1)(x+2)^2} = \frac{A}{x-1} + \frac{B}{x+2} + \frac{C}{(x+2)^2}$$

$$1 = A(x+2)^2 + B(x-1)(x+2) + C(x-1)$$

$$x = 1 \Rightarrow A = \tfrac{1}{9}$$

$$x = -2 \Rightarrow C = -\tfrac{1}{3}$$

$$x = 0 \Rightarrow 1 = \tfrac{4}{9} - 2B + \tfrac{1}{3} \Rightarrow B = -\tfrac{1}{9}$$

$$\therefore \frac{1}{(x-1)(x+2)^2} = \frac{1}{9}\left(\frac{1}{x-1} - \frac{1}{x+2} - \frac{3}{(x+2)^2}\right)$$

(b)

$$(x-1)\frac{dy}{dx} - y = \frac{x-1}{(x+2)^2}$$

$$\frac{dy}{dx} - \frac{1}{x-1}y = \frac{1}{(x+2)^2}$$

Integrating factor: $\exp\left(\int -\frac{1}{x-1}dx\right)$

$$= \exp(-\ln(x-1)) = (x-1)^{-1}$$

$$\frac{1}{(x-1)}\frac{dy}{dx} - \frac{1}{(x-1)^2}y = \frac{1}{(x-1)(x+2)^2}$$

$$\frac{d}{dx}\left(\frac{y}{x-1}\right) = \frac{1}{(x-1)(x+2)^2}$$

$$= \frac{1}{9}\left(\frac{1}{x-1} - \frac{1}{x+2} - \frac{3}{(x+2)^2}\right)$$

$$\frac{y}{x-1} = \frac{1}{9}\left(\ln|x-1| - \ln|x+2| + \frac{3}{x+2}\right) + c$$

$$y = \frac{x-1}{9}\left(\ln|x-1| - \ln|x+2| + \frac{3}{x+2}\right) + c(x-1)$$

$$= \frac{x-1}{9}\left(\ln\frac{|x-1|}{|x+2|} + \frac{3}{x+2}\right) + c(x-1)$$

16. (a) For $n = 1$, the LHS $= \cos\theta + i\sin\theta$ and the RHS $= \cos\theta + i\sin\theta$. Hence the result is true for $n = 1$.

Assume the result is true for $n = k$, i.e.
$(\cos\theta + i\sin\theta)^k = \cos k\theta + i\sin k\theta$.

Now consider the case when $n = k + 1$:
$(\cos\theta + i\sin\theta)^{k+1} = (\cos\theta + i\sin\theta)^k(\cos\theta + i\sin\theta)$
$= (\cos k\theta + i\sin k\theta)(\cos\theta + i\sin\theta)$
$= (\cos k\theta\cos\theta - \sin k\theta\sin\theta) + i(\sin k\theta\cos\theta + \cos k\theta\sin\theta)$
$= \cos(k+1)\theta + i\sin(k+1)\theta$
Thus, if the result is true for $n = k$ the result is true for $n = k + 1$.
Since it is true for $n = 1$, the result is true for all $n \geq 1$.

(b)

$$\frac{(\cos\frac{\pi}{18} + i\sin\frac{\pi}{18})^{11}}{(\cos\frac{\pi}{36} + i\sin\frac{\pi}{36})^4} = \frac{\cos\frac{11\pi}{18} + i\sin\frac{11\pi}{18}}{\cos\frac{\pi}{9} + i\sin\frac{\pi}{9}}$$

$$= \frac{\cos\frac{11\pi}{18} + i\sin\frac{11\pi}{18}}{\cos\frac{\pi}{9} + i\sin\frac{\pi}{9}} \times \frac{\cos\frac{\pi}{9} - i\sin\frac{\pi}{9}}{\cos\frac{\pi}{9} - i\sin\frac{\pi}{9}}$$

$$= \frac{\cos\frac{11\pi}{18}\cos\frac{\pi}{9} + \sin\frac{11\pi}{18}\sin\frac{\pi}{9}}{\cos^2\frac{\pi}{9} + \sin^2\frac{\pi}{9}} + \text{imaginary term}$$

$$= \cos\left(\frac{11\pi}{18} - \frac{\pi}{9}\right) + \text{imaginary term}$$

$$= \cos\frac{\pi}{2} + \text{imaginary term}$$

Thus the real part is zero as required.

Hey! I've done it